Practical Amateur Astronomy
## How to Use a Computerized Telescope

Computerized telescopes have brought a revolution to amateur astronomy. The new technology has opened up observing to many who were previously daunted by the task of learning the sky or using star charts. Finding an astronomical object becomes a quick operation with a computerized telescope, allowing more time for actual observation of the heavens.

*How to Use a Computerized Telescope* is the first handbook that describes how to get your computerized telescope up and running, and how to embark on a program of observation. It explains in detail how the sky moves, how your telescope tracks it, and how to get the most out of any computerized telescope. Packed full of practical advice and tips for troubleshooting, it translates the manufacturers' technical jargon into easy-to-follow, step-by-step instructions, as well as including many of the author's tried and tested observing techniques. Early chapters explain how to test your telescope's optics, choose eyepieces and accessories, take pictures through your telescope, and diagnose operational problems. The second half of the book then gives detailed instructions for three classic telescopes: the Meade LX200, Celestron NexStar 5 and 8, and Meade Autostar (ETX and LX90). Besides helping owners and would-be purchasers of these models, the instructions also provide a basis of comparison for understanding newer telescopes.

Amateur astronomers will find this book an invaluable source of information and advice for getting started with a new computerized telescope. Concentrating mainly on telescope operation and troubleshooting, it is the ideal companion to *Celestial Objects for Modern Telescopes*, also by Michael Covington, which provides the reader with suggestions for interesting celestial objects to view and advice on how to observe them.

MICHAEL COVINGTON, an avid amateur astronomer since age 12, has degrees in linguistics from Cambridge and Yale Universities. He does research on computer processing of human languages at the University of Georgia, where his work won first prize in the IBM Supercomputing Competition in 1990. His current research and consulting areas include theoretical linguistics, natural language processing, logic programming, and microcontrollers. Although a computational linguist by profession, he is recognized as one of America's leading amateur astronomers and is highly regarded in the field. He is the author of several books, including the highly acclaimed *Astrophotography for the Amateur* (1985; second edition 1999) and *Celestial Objects for Modern Telescopes* (2002), which are both published by Cambridge University Press. The author's other pursuits include amateur radio, electronics, computers, ancient languages and literatures, philosophy, theology, and church work. He lives in Athens, Georgia, U.S.A., with his wife Melody and daughters Cathy and Sharon, and can be visited on the Web at www.covingtoninnovations.com.

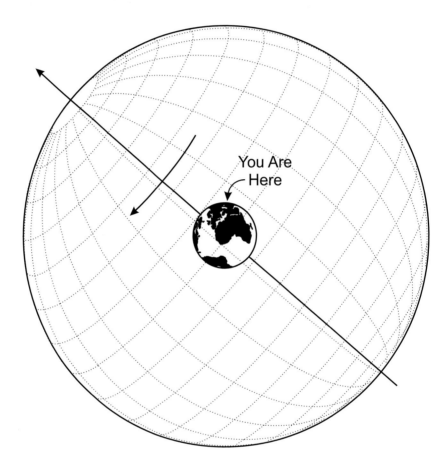

Practical Amateur Astronomy

# How to Use a Computerized Telescope

Michael A. Covington

CAMBRIDGE
UNIVERSITY PRESS

PUBLISHED BY THE PRESS SYNDICATE OF THE UNIVERSITY OF CAMBRIDGE
The Pitt Building, Trumpington Street, Cambridge, United Kingdom

CAMBRIDGE UNIVERSITY PRESS
The Edinburgh Building, Cambridge CB2 2RU, UK
40 West 20th Street, New York, NY 10011-4211, USA
477 Williamstown Road, Port Melbourne, VIC 3207, Australia
Ruiz de Alarcón 13, 28014 Madrid, Spain
Dock House, The Waterfront, Cape Town 8001, South Africa

http://www.cambridge.org

First published 2002

Printed in the United Kingdom at the University Press, Cambridge

*Typefaces* Palatino 10/13 pt and Meta Plus book      *System* LATEX $2_\varepsilon$   [TB]

*A catalogue record for this book is available from the British Library*

*Library of Congress Cataloguing in Publication data*

Covington, Michael A., 1957–
How to use a computerized telescope / Michael A. Covington.
     p.     cm.
Includes bibliographical references and index.
ISBN 0 521 00790 9 (pbk.)
1. Telescopes – Automatic control – Handbooks, manuals, etc.   2. Computerized
instruments – Handbooks, manuals, etc.   I. Title.
QB88 .C79   2002
522'.2 – dc21   2002024693

ISBN 0 521 00790 9 paperback

Soli Deo gloria

# Contents

# Preface

Computerized telescopes are revolutionizing amateur astronomy. Even the least expensive entry-level telescopes are now available with computer-controlled motors to find and track objects in the sky. No longer do you have to search for NGC 1999 or Neptune by carefully comparing the view with a star map – you just tell the telescope what to point at, and it does it.

Do computer controls take all the fun out of astronomy? No more than paved highways take the fun out of the Arizona desert. Professional astronomers have used setting circles to find objects since the time of Tycho Brahe and have always tried to make them as accurate as possible. Amateurs have long *had* setting circles, but they weren't very accurate. Now, with the advent of computers, professional-level accuracy is within the reach of the amateur, and the computer actually controls the telescope rather than just telling you where it's pointed.

After 30 years of finding celestial objects the old way, I bought my first computerized telescope in 2000 and immediately found myself doing a new kind of amateur astronomy. Suddenly I was spending my time looking *at* objects instead of *for* them. No longer preoccupied with "star-hopping", I could spare the time and attention to study the celestial objects themselves.

In fact I realized for the first time that, for all those years, my observing program had been skewed by the fact that some objects are easier to find than others. I regularly viewed M13 and not M92 because the latter is not near any bright stars. I rarely looked at Uranus or Neptune because that would require getting out a special map, updated yearly. Now I can look at anything within reach of the telescope.

At the same time I have become much more aware of, and dependent on, astronomical data sources. If an atlas omits NGC 404 or a star catalogue skips ξ Ursae Majoris, that's an obstacle I'll bump into and notice. If I use epoch-1950 coordinates on an epoch-2000 telescope, I won't find what I'm looking for. Conversely, the latest data files from the Astronomical Data Center can be put to immediate use with my computer and telescope.

What it all means is that new-style amateur astronomers need a new kind of guidebook. My writing project began as a list of interesting objects that I put together for use at the telescope. Soon I added a concise summary of the Meade LX200 operating manual. Simon Mitton of Cambridge University Press saw my notes and encouraged me to turn them into a book. By the time I finished, I had enough material for two books, *How to Use a Computerized Telescope* (this volume) and *Celestial Objects for Modern Telescopes* (the companion volume, which focuses on the sky rather than the equipment).

While I was writing the two books, Scott Roberts of Meade Instruments lent me equipment to try out. The technical support departments at Meade, Celestron, Software Bisque, and Starry Night Software answered technical questions. Daniel Bisque supplied software for testing. Howard Lester, Dennis Persyk, Lenny Abbey, Rich Jakiel, T. Wesley Erickson, Robert Leyland, R. A. Greiner, Richard Seymour, Ralph Pass, Phil Chambers, Ells Dutton, Michael Forsyth, and John Barnes critiqued drafts of parts of the text. Tom Sanford let me try out his Meade LX90 at length. Earlier, Jim Dillard first got me interested in computer-aided astronomy by buying my old Meade LX3 from me and outfitting it with digital setting circles. There are probably others whose names I've forgotten to list, and I beg their indulgence. And I have hopelessly lost track of who helped with which volume!

All along, Melody (my wife) and Cathy and Sharon (my daughters) have patiently endured a living room full of tripods and have even accompanied me on some observing trips. (I keep pointing out that all this is not as expensive as boats or even golf!) I want to thank all of these people, and others unnamed, for their encouragement and assistance.

Please visit me on the Web at http://www.covingtoninnovations.com, where this book will have its own web page with updates and related information.

*Athens, Georgia*
*December 24, 2001*

# Part I
# Telescopes in general

# Chapter 1
# Welcome to amateur astronomy!

Welcome to amateur astronomy! If you are new to this field, and especially if you have never owned a telescope before, this chapter is for you. Otherwise, feel free to skip ahead. I've tried to write a book that I'll actually use while observing. Parts of it are quite specialized; take what suits you and save the rest for later.

Amateur astronomy, like other hobbies, is something you can go for a little or a lot. Computerized telescopes make casual stargazing easier than ever before, since you don't have to gather up star maps and look up planet positions before going out under the sky. At the other end of the spectrum, the advanced amateur with a busy, semi-professional observing program will find that a computerized telescope is a real time-saver. Both approaches to amateur astronomy are respectable, and so is everything in between.

The key to enjoyment is to have realistic expectations and continue building your knowledge and skill. Looking through a telescope is a very different experience from looking at photographs in books, and it may take some getting used to. If you don't already have a telescope, get some experience looking through other people's telescopes before buying one of your own. Contact a local astronomy club if possible.

## 1.1 Using a telescope

Newcomers are sometimes surprised to find that spectacular objects such as the Horsehead Nebula are not normally visible in telescopes at all – the eye cannot accumulate light the way the camera does. But other nebulae, such as M42, are far more spectacular visually than on film because the eye can see the faint outer regions without overexposing the center. Likewise, the three-dimensional ball shape of a globular star cluster is more impressive "live" than in pictures because the eye covers a greater brightness range. And the ever-changing phenomena of Jupiter, Saturn, sunspots, and variable stars provide a constant supply of new sights – though the real colors of the planets are much more subtle than the bright colors of computer-processed pictures.

One important tactic is to use low power. Unlike microscopes, telescopes do not perform well at their highest powers; this is true of *all* telescopes because of the wave properties of light and the turbulence of the Earth's atmosphere. Most astronomy is done at 20× to 100×, with 15- to 40-mm eyepieces. Use whatever eyepiece gives the most comfortable view – usually the lowest-power one – and switch to high power only when actually necessary.

## 1.2    Learning the sky

A computerized telescope will help you learn the sky; it won't eliminate the need to do so. Every time you set up your telescope, you will need to identify at least two bright stars. Although the telescope will try to find the stars for you, things go much more smoothly if you learn to recognize them on your own.

Don't try to memorize a star map; that would be tiresome. Instead, find something in the sky that catches your eye, then use a map to identify it. (My personal career began with the Belt of Orion.) Some constellations, such as Ursa Major and Cassiopeia, jump right out at you; others are obscure, and you will never need to learn them. Not one astronomer in a hundred can sketch Camelopardalis from memory.

You'll also need to build your awareness of how the sky moves, how the moon goes through its phases, and so forth. That's what Chapter 2 is about, but a couple of hours of *watching*, repeated every week, will make the sky come alive in a way that no diagram can do.

Above all, though, don't let imperfect conditions, imperfect equipment, or a lack of technical mastery keep you from looking at the sky. On the first night with a new telescope, just take it outside and *look!* Start by viewing distant treetops and the Moon; then examine anything that looks interesting – bright stars, star clusters, the Milky Way, or whatever you can see, whether or not you can identify it. You've just begun a lifelong adventure.

## 1.3    Is a computerized telescope right for you?

Computerized telescopes are not ideal for everyone. There are three situations in which I do not recommend purchasing one.

First, if you are completely unfamiliar with the sky, you are probably not ready for any telescope, not even a computerized one. Instead, get some star maps and perhaps a pair of binoculars, and spend several evenings looking at the stars, learning some constellations, and becoming aware of how the sky moves. Until you can identify at least a few constellations, you will not be able to set up a computerized telescope reliably, nor will you know whether it's working correctly once you get it going.

Second, if you need maximum optical performance at minimum cost, you will probably not want to spend the extra money for a computerized mount. Instead, go for a **Dobsonian** telescope (a Newtonian on a low-cost altazimuth mount that looks like a cannon). Finding objects with a Dobsonian takes considerable skill and extensive use of star maps, but the views are rewarding, especially under a dark country sky. A given amount of money will buy a much larger Dobsonian than any other kind of telescope. Dobsonians can be outfitted with computerized motor drives later.

Third, if you want to do astrophotography on a limited budget, you need smooth drive motors, which the cheapest computerized telescopes do not have. Save up for a high-end computerized telescope that is specifically designed for astrophotography, or stick with conventional AC motors.

## 1.4 Material you can skip

Smaller type like this indicates technical material that you can skip until you need it. By printing it in smaller type, I avoided having to take it out of its logical place in the text.

## 1.5 Does this book cover your telescope?

All computerized telescopes work on the same basic principles. Chapters 10–12 give detailed instructions for the Meade LX200, Celestron NexStar and Meade Autostar (ETX and LX90), based on my experience with three specific telescopes purchased in 2000 and 2001.

By the time you read this, none of these telescopes will be the current model. As manufacturers continue to update their software and expand their product lines, the information in those chapters will inevitably become out of date. But plenty of older telescopes remain in use, and in general, the oldest versions of the software are the most in need of explanation. Newer versions have better documentation, and the older versions are always a good starting point for comparison.

# Chapter 2
# How the sky moves

## 2.1 Daily motion

The universe is a mass of swirling motions, but most of the time, you can ignore all but one of them. That one is **daily motion** (**diurnal motion**), caused by the rotation of the Earth. You will see it immediately if you aim a $100\times$ telescope at a star with the drive motor turned off.

As you know, celestial objects rise in the east, move across the sky, and set in the west. But as Figures 2.1 and 2.2 show, that is not the whole story. The motion is not directly from east to west; instead the whole sky rotates like a globe with Polaris at its north pole.

In the southern sky, each object rises somewhere on the eastern horizon (not necessarily due east!), passes across the sky, and sets somewhere in the west. Its path may be long or short. In the far southern sky, objects rise just east of south, climb only a short distance above the horizon, and set again a short time later, just west of south.

> *Hint:* Maps of the sky have north at the top, east at the *left* (not right as on a terrestrial map), in order to match the view that you see when facing south and looking up. Get used to facing south to get your bearings when looking at the sky.

The most northerly celestial objects are **circumpolar**; that is, they do not rise or set at all. Instead, they twirl around the north celestial pole, which is conveniently marked by the star **Polaris**. Each revolution takes one day (24 hours).

Opposite the circumpolar region, there is a region in the far south containing stars that never rise. That is why Alpha Centauri, for instance, is not visible from the continental United States.

The diagrams show the sky as seen from New York. Farther north, everything in the south is lower, everything in the north is higher, and the circumpolar region is larger. Within the Arctic Circle, the Sun can become circumpolar; that's how

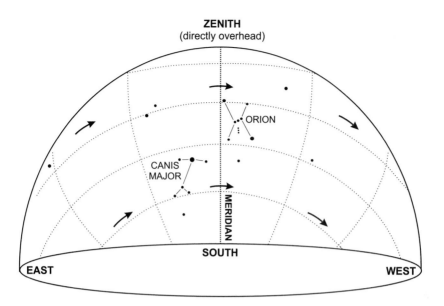

Figure 2.1. The southern sky at the latitude of New York (40° N) at 8 p.m. on February 21.

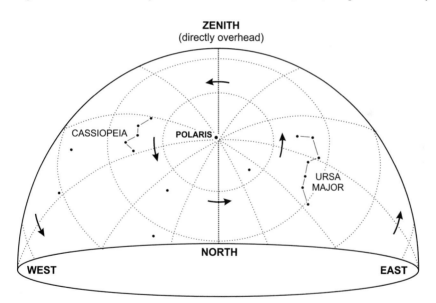

Figure 2.2. The same, but looking north. Celestial objects twirl around Polaris once every 24 hours.

they get the Midnight Sun. Even in England, the Sun is so nearly circumpolar in mid-June that the sky does not get completely dark.

At more southerly latitudes, the opposite is the case. From Florida, you can see the star Canopus, which is due south of Sirius and below the New York horizon.

Seen from the equator, Polaris lies on the northern horizon, and nothing is circumpolar, but nothing is hidden from view in the south; you can see the entire sky. From the southern hemisphere (Australia, for instance), Polaris is below the horizon but the south celestial pole is high in the sky; from there, you can see all of the southern sky but not all of the north.

## 2.2 Coordinates

### 2.2.1 Right ascension and declination

The alert reader will notice that Figures 2.1 and 2.2 are crisscrossed by lines that look like latitude and longitude on a globe.

That's exactly what they are, but the globe to which they refer is the **celestial sphere** (Figure 2.3). This is an imaginary sphere, infinitely large, surrounding the Earth, on which the stars have fixed positions. Since the sphere is imaginary, its motion can be imaginary too, so astronomers pretend that the Earth holds still (with the observer's location right on top, of course) and the sphere rotates around it. Its rotation is that of the Earth, but in the opposite direction.

Coordinates on the celestial sphere are called **declination** (abbreviated **Dec.** or $\delta$) and **right ascension** (**R.A.**, **AR**, or $\alpha$). Together, right ascension and declination are known as **equatorial coordinates** because they are based on a sphere whose equator and poles correspond to those of the Earth.

Declination is like latitude and is measured in degrees, negative if south of the equator. Right ascension is like longitude but is measured in hours (0 to 24). Since the sphere rotates once per day, it makes sense to measure longitude in time units.

Figure 2.4 shows the whole sky (as seen at the same time and place as Figures 2.1 and 2.2) with lines of R.A. and declination labeled.

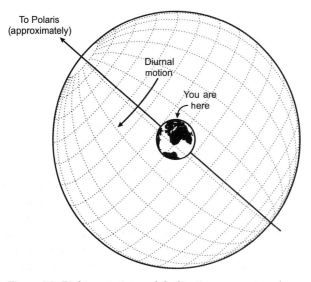

Figure 2.3. Right ascension and declination are measured on an imaginary sphere that surrounds the Earth, on which the stars have fixed positions.

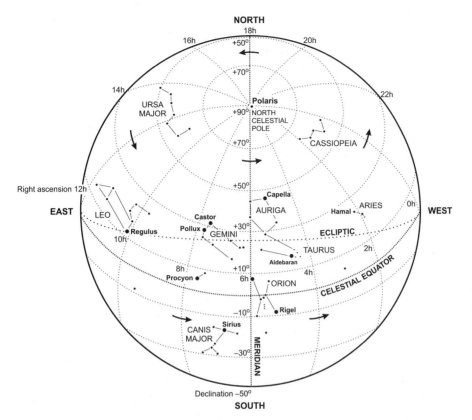

Figure 2.4. The whole sky as seen in Figures 2.1 and 2.2. Based on a chart created with *Starry Night Pro* astronomy software (http://www.starrynight.com), reproduced by permission.

> *Hint:* In astronomy, the directions *north, south, east*, and *west* always refer to the celestial sphere unless otherwise indicated. To move north in the sky means to move toward the north celestial pole.

The right ascensions and declinations of the stars are, for all practical purposes, constant. The Sun, Moon, and planets move around on the celestial sphere, so their right ascensions and declinations vary.

The term *right ascension* sounds as if it refers to something ascending or rising at a right angle, and indeed it does. Seen from the Earth's equator, all celestial objects rise and set at right angles to the horizon, at times that depend directly on their right ascensions. *Declination* is from the Latin word for "bending", an appropriate name for an angle.

## 2.2.2 Declination and latitude

The altitude of the celestial pole above the horizon equals the observer's latitude. Thus, at New York (latitude $+40°$), the pole is $40°$ above the horizon.

The declination of a star directly overhead also equals the observer's latitude. Thus, $\epsilon$ Persei (declination $+40°$) can pass directly overhead at New York, but the Sun (which is never north of $+23.4°$) cannot.

Letting $L$ stand for the observer's latitude, a star is circumpolar if its declination is north of $90°-L$, and never rises if it is south of $L-90°$. Thus, at New York, stars north of declination $+50°$ are circumpolar, and those south of $-50°$ never rise.

### 2.2.3 Some terminology

As shown in Figures 2.1–2.2, the **zenith** is the point directly overhead, and the **meridian** is the line that runs directly north and south through the zenith.

**Altitude** is height above the horizon, measured as an angle. The horizon is at altitude $0°$; the zenith is at $90°$.

**Azimuth** is direction measured *rightward from north* ($0°$) through east ($90°$), south ($180°$), west ($270°$), and back to north ($360°$).

However, Meade LX200 telescopes follow an older tradition and reckon azimuth rightward from south through west, giving values that are $180°$ away from ordinary azimuths.

Together, altitude and azimuth are known as **horizontal coordinates** since they use the horizon as their equator. Computerized telescopes compute the altitude and azimuth of celestial objects automatically from the right ascension and declination.

Right ascension can be measured in degrees ($1^h = 15°$; $1° = 4$ minutes of R.A.). When measured this way, it is often called **sidereal hour angle** (**SHA**).

The **local hour angle** (or just **hour angle**, abbreviated **HA**) of a star is the difference between its right ascension and that of the meridian. For example, in Figure 2.4, the hour angle of Rigel is almost 1 hour west, or $+15°$.

Telescope instructions often say to aim the telescope at **hour angle zero** during setup. That means to aim it at the meridian, i.e., due south.

### 2.2.4 Other coordinate systems

Besides right ascension and declination, two other systems of fixed coordinates are used in the sky.

The **ecliptic**, marked on Figure 2.4, is a line that indicates the plane of the Earth's orbit. The Sun is always on the ecliptic, and the Moon and planets are always close to it. **Ecliptic latitude** and **ecliptic longitude**, often called just **latitude** and **longitude**, use the ecliptic in place of the celestial equator. Ecliptic coordinates are used in calculating planetary orbits but not, nowadays, for much else, though in ancient times they were the basis of star catalogues and maps. Like right ascension, ecliptic longitude starts at $0°$ at the point where the Sun crosses the celestial equator in the spring.

**Galactic latitude** and **galactic longitude** are measured relative to the plane of our galaxy, with longitude $0°$ defined as the galactic center in Sagittarius. These coordinates are not used for finding objects, but they are important to astrophysicists because they provide a quick way to tell what part of our galaxy we are looking at (or through) when we view an object.

The Greek letters beta ($\beta$) and lambda ($\lambda$) stand for latitude and longitude, respectively, in these and other systems.

### 2.2.5 Degrees, minutes, and seconds

Because the celestial sphere appears to be infinitely large, distances on it can only be measured as angles. For example, the apparent width of the Moon is half a degree, which means that two opposite edges of the Moon make a 0.5° angle with its vertex at the observer's eye (Figure 2.5). In similar terms, we can talk about two stars being ten degrees apart or the true field of a telescope being a quarter of a degree.

There are 360 **degrees** (360°) in a full circle. Each degree is divided into 60 **arc-minutes** or simply **minutes** (60′), and each minute is divided into 60 **arc-seconds** or simply **seconds** (60″). Thus $1' = 1/60°$ and $1'' = 1/3600°$.

Some calculators have a built-in function to convert degrees, minutes, and seconds into decimal degrees and vice versa. If yours doesn't, the following examples will show how the conversion is done.

To convert 33°45′18″ into decimal degrees, simply add up its component parts:

$$33°45'18'' = 33 + \frac{45}{60} + \frac{18}{3600} = 33 + 0.75 + 0.005 = 33.755°$$

If the angle is negative, the minutes and seconds are included within the negation:

$$-33°45'18'' = -\left(33 + \frac{45}{60} + \frac{18}{3600}\right) = -(33 + 0.75 + 0.005) = -33.755°$$

It is probably easiest to ignore the minus sign until after the conversion.

Converting the other way is more complicated. First split the integer part from the fractional part:

$$33.755° = 33° + 0.755°$$

Now convert the fractional part into minutes:

$$0.755° = (0.755 \times 60)' = 45.3'$$

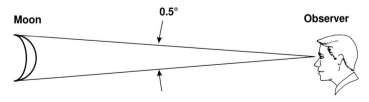

Figure 2.5. The apparent width of the Moon is half a degree. (From *Astrophotography for the Amateur*, Cambridge, 1999.)

11

If decimal minutes are good enough, you're done. Otherwise convert the fractional part of the minutes into seconds:

$$45.3' = 45' + 0.3'$$

$$0.3' = (0.3 \times 60)'' = 18''$$

Thus $33.755° = 33°45'18''$.

### 2.2.6 Distance between points in the sky

In trigonometry and on pocket calculators, angles are sometimes measured in **radians** (rad), where:

$$1 \text{ rad} = \frac{180°}{\pi} \approx 57.29578° \approx 3438' \approx 206\,265''$$

The angular size of a celestial object in radians is its true size divided by its distance. Thus a pair of stars 2000 light-years away and 2 light-years apart will have an apparent separation of 1/1000 radian.[1]

The angular distance between two celestial objects is not the same as the difference of their right ascensions and declinations. Near the poles, the lines of right ascension come together, and very close neighbors can have very different right ascensions. The formula for angular distance between two objects of known coordinates is:

$$\text{Distance} = \arccos[\sin \delta_1 \sin \delta_2 + \cos \delta_1 \cos \delta_2 \cos(\alpha_1 - \alpha_2)]$$

where $\delta_1, \delta_2$ are their respective declinations and $\alpha_1, \alpha_2$ are the right ascensions. Remember to express right ascension as an angle (1 hour = 15°). Other types of latitude and longitude (e.g., ecliptic or galactic coordinates, or even altitude and azimuth) can be used in place of declination and right ascension in the formula.

## 2.3 Annual motion

### 2.3.1 Why time of year matters

The maps in Figures 2.1–2.4 are drawn for 8 p.m. on February 21. If you compare these maps to the sky a month later – at 8 p.m. on March 21 – you'll see that everything has shifted westward. Castor and Pollux will be west of the meridian, Leo will be higher than the map shows it, and Aries will be about to set.

This shift is called **annual motion** and is caused by the yearly revolution of the Earth around the Sun. It amounts to 2 hours of right ascension per month, or 24 hours per year, or 4 minutes per day. To get a feel for it, look at the whole-sky charts for successive months in an astronomy magazine or a book such as *Celestial Objects for Modern Telescopes*, the companion to this volume.

---

[1] Strictly speaking, this is true only for objects that are curved to match the celestial sphere – and none of them really are. It is accurate for objects of all shapes when the angles are small.

One way to understand annual motion is to look at the sky at midnight every night of the year. Naturally, at midnight, your meridian will be directly opposite the Sun; that's what "midnight" means. But "directly opposite the Sun" is a direction that depends on the position of the Earth in its orbit. That's why the stars will seem to shift, completing a full circle in one year.

Another approach is to think in terms of the Sun's position in the sky. (Although we cannot see stars in the daytime, the Sun has a right ascension and declination just like everything else; devising ways to measure its position was one of the main problems of ancient astronomy.) The Sun moves eastward along the ecliptic, increasing its right ascension about 4 minutes per day. But we keep time by the Sun, not the stars, so the effect is that the stars seem to be moving westward.

A simple revolving map called a **planisphere** (Figure 2.6) can help you keep track of annual motion. So can computer programs such as *Starry Night*, *TheSky*, and *SkyMap Pro* (see p. 36).

### 2.3.2 Sidereal time

The right ascension of the meridian is called the **sidereal time** and is, of course, constantly changing. A sidereal clock runs slightly faster than an ordinary clock; it gains about 4 minutes per day, or exactly one day per **sidereal year** (365.2564 days), because the Earth's orbit around the Sun, over the course of a year, is equivalent to one extra rotation about its axis.

Sidereal time is useful for planning observing sessions because objects are highest in the sky when their R.A. is close to the sidereal time. Here is an approximate formula for sidereal time that is simple enough to calculate in your head:

Sidereal time (in hours) $\approx$ hours past noon $+\, 2 \times$ (months past March 21)

If the result is greater than 24, subtract 24.

For example, at 9 p.m. on June 7, it is 9 hours past noon and 2.5 months past March 21. Accordingly, the sidereal time is $9 + (2 \times 2.5) = 13$ hours.

March 21 is the date of the **vernal equinox**, when the Sun crosses the celestial equator. Julius Caesar's advisors invented leap years in order to keep the civil calendar in sync with this event, and hence with the seasons. The period between vernal equinoxes is a **tropical year**, 365.2422 days, not quite equal to a sidereal year because of precession (see p. 17).

## 2.4 Time of day

### 2.4.1 Solar time and time zones

We normally keep time by the Sun, not the stars. That is, ordinary human time-keeping is based on **mean solar time**. Here "mean" means that we use a clock

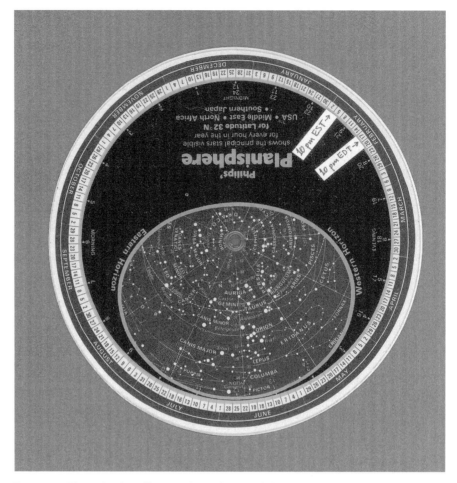

Figure 2.6. The author's well-worn planisphere, with handmade labels showing the difference between local mean time and zone time (winter and summer). Local horizon could also be drawn in.

that runs at a constant speed. The actual Sun is slightly ahead of its mean position at some times of year, and slightly behind at other times, because the Earth's orbit is not a perfect circle.

The mean solar time at longitude 0° (Greenwich, England) is called **Greenwich Mean Time (GMT)**, **Universal Time (UT)**, or sometimes, in military parlance, **Z time** or **Zulu time**.

The Earth is divided into **time zones** about an hour wide, each of which uses the mean solar time of a particular longitude. For example, the entire U.S. Eastern time zone uses the mean solar time for longitude 75° west. Table 2.1 describes the time zones of Britain, Canada, and the United States. For the rest of the world, see Figure 2.7.

The difference between standard time and local mean solar time is exactly 4 minutes per degree of longitude. For example, in Atlanta, at longitude

Table 2.1 *Time zones of Britain, Canada, and the United States*

| Britain | 0° | Greenwich Mean Time = UTC |
|---|---|---|
| | | British Summer Time = UTC+1$^h$ |
| Newfoundland | 52.5° W | Newfoundland Standard Time = UTC−3$^h$30$^m$ |
| | | Newfoundland Daylight Time = UTC−2$^h$30$^m$ |
| Atlantic | 60° W | Atlantic Standard Time = UTC−4$^h$ |
| | | Atlantic Daylight Time = UTC−3$^h$ |
| Eastern | 75° W | Eastern Standard Time = UTC−5$^h$ |
| | | Eastern Daylight Time = UTC−4$^h$ |
| Central | 90° W | Central Standard Time = UTC−6$^h$ |
| | | Central Daylight Time = UTC−5$^h$ |
| Mountain | 105° W | Mountain Standard Time = UTC−7$^h$ |
| | | Mountain Daylight Time = UTC−6$^h$ |
| Pacific | 120° W | Pacific Standard Time = UTC−8$^h$ |
| | | Pacific Daylight Time = UTC−7$^h$ |
| Alaska | 135° W | Alaska Standard Time = UTC−9$^h$ |
| | | Alaska Daylight Time = UTC−8$^h$ |
| Hawaii | 150° W | Hawaii Standard Time = UTC−10$^h$ |

84° west, the local mean solar time is 36 minutes behind Eastern Standard Time because Atlanta is 9° west of the longitude on which the time zone is based.

Most localities observe **daylight saving time** (**summer time**) by putting their clocks ahead one hour in the summer. This can be interpreted as an attempt to hold sunrise nearer a fixed time of day, so that the increased length of summer days shows up mainly in the evening.

Time zones UTC+12 and UTC−12 are 24 hours apart; they meet at the International Date Line in the Pacific Ocean. They ought to be the beginning and end of the whole system, but in fact there is also a zone UTC+13 (New Zealand Summer Time), and UTC+14 is used in parts of the Kiribati Republic.

## 2.4.2 Hints on using UT

Because UT is 5 hours ahead of Eastern Standard Time, UT midnight occurs at 7 p.m. in the eastern United States, and the UT date changes in the middle of an early-evening observing session. Farther west, the UT date is ahead of the local date throughout the evening. I once missed a lunar eclipse because when converting the time from UT, I forgot to convert the date.

You can get the exact time, in UT, from any GPS receiver, or by listening to shortwave radio station WWV (2.5, 5.0, 10.0, and 15.0 MHz), or from Internet sites such as http://www.usno.navy.mil.

# WORLD MAP OF TIME ZONES

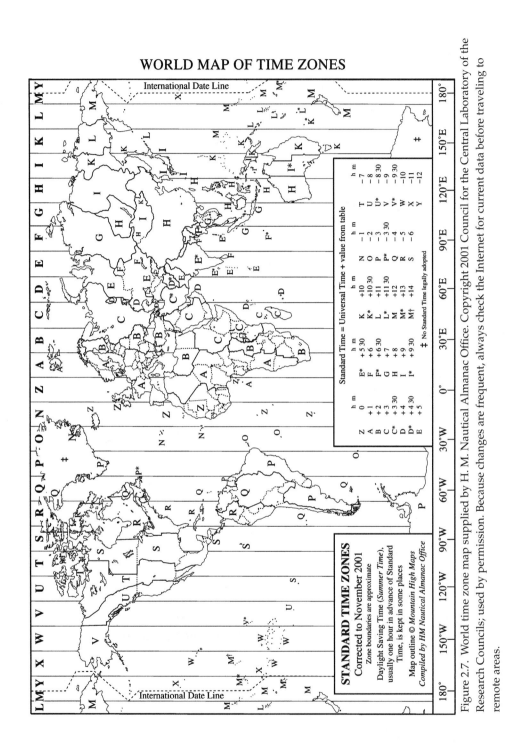

Figure 2.7. World time zone map supplied by H. M. Nautical Almanac Office. Copyright 2001 Council for the Central Laboratory of the Research Councils; used by permission. Because changes are frequent, always check the Internet for current data before traveling to remote areas.

### 2.4.3 UTC, ET, and other subtleties

The official standard for UT is maintained by several national bureaus of standards and is called **Coordinated Universal Time** or **UTC**.

UTC is determined by an atomic clock whose speed is more uniform than the rotation of the Earth. Because the Earth is gradually slowing down, UTC must sometimes insert a **leap second** to keep the clock in sync with the Earth's rotation. There has been a leap second at the end of almost every year since 1972.

Because nobody can predict when the authorities will decide to insert a leap second, UTC cannot be used for precise astronomical predictions. Instead, predictions of planetary positions, eclipses, and the like are expressed in **ephemeris time (ET)** or **dynamical time (TT, TDT, TDB)**. These timescales are perfectly smooth but gradually get out of step with the Earth's rotation.

As of 2001, ET, TT (TDT), and TDB are almost exactly equal and are all about 60 seconds ahead of UTC. The difference between ET and UTC is called **ΔT**.

## 2.5 Slow changes in R.A. and declination

### 2.5.1 Precession and epochs

The right ascensions and declinations of stars change slowly over the years because of *precession*, a gradual shift in the direction of the Earth's axis.

It follows that right ascension and declination are meaningless unless you know the **epoch** (the date) to which they apply. Most star maps and catalogues now use epoch 2000, or more precisely **J2000.0** (noon UTC, January 1, 2000, twelve hours into the new year). Many earlier maps use epoch 1950.

A rough way to convert coordinates from 1950 to 2000 is to *add $2\frac{1}{2}$ minutes to the right ascension* and ignore the change in declination. This is far from exact, but it will get you to within half a degree of the right position anywhere in the sky, and in most parts of the sky it's much better than that. It's a quick way to aim an epoch-2000 telescope when all you have are epoch-1950 coordinates. Exact conversion tables and formulae are given in *Celestial Objects for Modern Telescopes* and other handbooks.

Many listings of planet positions, including the *Astronomical Almanac*, use the **epoch of current date** – that is, the epoch is the date for which the predictions are made. Computerized telescopes, however, invariably use epoch-2000 coordinates. As this is written, in 2002, the two are very close.

### 2.5.2 How precession works

Right ascension zero is defined as the point where the Sun crosses the celestial equator, moving northward, in the spring. Because of precession, this point is not fixed. Its movement can be visualized in two ways.

One way to describe precession is to say that the north celestial pole is gradually moving among the stars, tracing out a large circle every 25 770 years. Right now the pole is close to Polaris; in 2100 it will be closer. In ancient Egyptian times the pole was close to the star Thuban ($\alpha$ Draconis). In 14 000 AD, Vega will be the pole star, and in 28 000 AD, Polaris will be the pole star again.

Another way to describe precession – absolutely equivalent, though it may not seem so – is to say that the whole sky is constantly slipping eastward along the ecliptic at the rate of about 1° per 72 years. Thus, all the stars are constantly gaining right ascension, except for small regions near the celestial pole. Stars move north or south in declination depending on their position.

Precession is subject to a slight wobble called **nutation**, and star positions also undergo a slight, cyclical shift due to the Earth's orbital velocity relative to the incoming light rays (**aberration of starlight**). Neither of these has any effect on amateur observations or on the measurement of positions by comparing to other nearby stars.

### 2.5.3 Proper motion

Some stars are close enough to the Earth that their motion through space is visible. These stars are said to have appreciable **proper motion** (which means "motion of their own", with *proper* as in *property*).

Barnard's Star, a ninth-magnitude red dwarf in Ophiuchus, is the star with the largest known proper motion, about $+10''$ in declination and $-0.05$ second in right ascension per year. It is moving nearly due north.

Only eight stars have proper motions greater than $5''$ per year. In general, proper motion is of no concern to users of computerized telescopes. Even Barnard's Star takes 351 years to move one degree.

# Chapter 3
# How telescopes track the stars

## 3.1 What's inside a computerized telescope

A computerized (or "go to") telescope is one that finds celestial objects by itself. Well, not exactly by itself – you have to show it the positions of two stars, and from there it can find everything else. Two stars are sufficient to determine the position of the whole celestial sphere.

The telescope takes two kinds of commands. You can tell it to **go to** a particular object, based on its current knowledge of the position of the sky, or you can tell it to **sync** (synchronize) on an object that you have identified and centered in the field. The latter is how you tell the telescope the exact position of the sky.

Besides all the parts of an ordinary telescope, a computerized telescope has a **computer**, **motors**, and **encoders**.

### 3.1.1 Computer

The computer translates right ascension and declination to the position of the telescope on its mount – a coordinate transformation that involves lots of spherical trigonometry (see p. 35).

The software built into the computer is called **firmware** and contains a built-in catalogue of stars and deep-sky objects, plus algorithms to compute the position of the Moon and planets, so that you can choose objects by name.

Computerized telescopes are a testimony to the low price of powerful computers. The Celestron NexStar 5, for example, has four CPUs in an internal network. Comparable computing power would have cost millions of dollars in the 1960s and would have filled several rooms. What's more, the telescope can be connected to an external computer – typically a laptop PC – for even more brainpower.

Figure 3.1. The author's Meade LX200 telescope, ready for action on an altazimuth mount.

### 3.1.2 Motors

The motors move the telescope for **slewing** (pointing at different objects) and **tracking** (keeping up with diurnal motion). These motors have to work smoothly over a wide range of speeds, from slewing at 6° per second to tracking at perhaps 0.004° per second.

Some **backlash** (slack in the gears) is bound to be evident at some speeds. Backlash shows up as a delay in slewing when you have changed direction (e.g., slewing south immediately after slewing north). Most telescopes offer adjustable compensation to reduce this delay; the computer stores numbers that tell it to speed up the motor briefly when backlash is expected. Too much compensation causes the opposite problem – movements begin with a sudden jerk.

With a light-duty mount you may even see a small amount of **runout** (movement perpendicular to the direction intended). Don't panic; it won't interfere seriously with use of the telescope.

Slewing is subject to **cord wrap limits** so that the telescope won't turn around and around, tangling its cords (including internal wiring). That's why, to go from south to west, the telescope may decide to swivel around through east and north. These limits are not fixed but depend on the telescope's recent movements. When running on internal batteries, the Celestron NexStar 5 has no cord wrap limits. Nor do Vixen mounts that use rotary contacts.

### 3.1.3 Encoders

The encoders are rotation sensors; they tell the computer about the telescope's movements, completing a closed-loop feedback system.

Encoders only measure *relative* motion. That is, they tell the computer how far the telescope has moved since it was powered up. They do not tell it the absolute position of the tube. That is one reason the telescope must be aligned on the stars every time it is turned on. The other reason is of course that the computer doesn't know the position of the base or tripod.

In most telescopes, the encoders respond only to movements made electrically; if you unlock the brakes and move the telescope by hand, the computer loses track of its position. The Celestron Ultima 2000, however, can be moved by hand without disrupting computerized operation.

If the encoders fail or become disconnected, but the motors still work, the telescope undergoes **runaway** – once it starts moving, it doesn't stop.

### 3.1.4 Digital setting circles (DSC)

Digital setting circles consist of an encoder and computer, but no motors, retrofitted to an older telescope. To find an object, you key it into the computer, but you supply the muscle power for slewing; the computer tells you which way to move and when to stop. The computer takes the place of the graduated circles (p. 53) that would otherwise give you an approximate readout of right ascension and declination.

## 3.2 Altazimuth and equatorial mounts

An **altazimuth** (altitude–azimuth) telescope mount swivels around an axis that points straight up, so that it moves up, down, left, and right. An **equatorial** mount has its main axis parallel to the axis of the Earth, so that it moves north, south, east, and west. Most fork-mounted telescopes can be set up either way (Figure 3.2). A telescope without a mount is called an **optical tube assembly** (**OTA**).

Before telescopes were computerized, equatorial mounts were the only kind that could track the stars. All an equatorial mount has to do is rotate around its polar axis at the right speed, and as far as the telescope is concerned, the stars

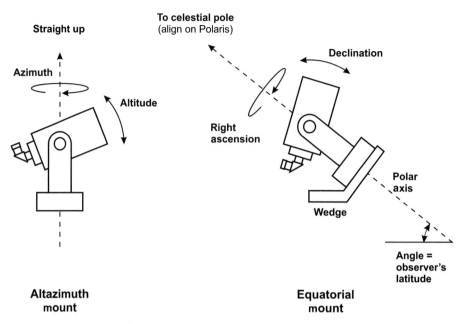

Figure 3.2. An equatorial mount is just an altazimuth mount tilted so that its main axis is parallel to that of the Earth.

stand still; any star that is in the field of view will stay there. Diurnal motion is completely overcome.

A computerized telescope can also track the stars in altazimuth mode, using motors on both axes, controlled by a computer that translates the Earth's rotation into altitude and azimuth. For this to be possible, the computer has to know the position of the celestial sphere relative to the earth, and that requires identifying (or in some cases guessing) the precise positions of two stars.

## 3.3 Site information

### 3.3.1 Why it's needed

Strictly speaking, a computerized telescope does not have to know the date, time, or location. As long as you can identify two stars, the telescope can find the rest of the celestial sphere without any further information. Indeed, if you power up a NexStar 5 and choose two-star alignment, it will not ask you the date, time, or site. Likewise, the Meade LX200 offers "unknown site" as an alignment option.

Generally, though, the telescope asks for the date, time, time zone, latitude, and longitude for several reasons:

- To suggest alignment stars and slew to their approximate positions at the beginning of the alignment process;
- To compute the exact position of the zenith (the point directly overhead) when doing "one-star" alignment;

- To warn you when the object you want to see is not in the sky;
- To find the positions of the Moon and planets in their orbits;
- To correct for atmospheric refraction, which slightly shifts objects that are low in the sky;
- To correct for parallax between different places on Earth when viewing the Moon;
- To correct for precession (which will be significant by 2010, though it is not noticeable as of 2002, relative to epoch 2000.0).

Not all telescopes do all these things. For instance, the original model NexStar 5 does not warn you when objects are not in the sky, does not correct for atmospheric refraction, and does not correct for parallax when viewing the Moon. The two-star alignment process itself takes care of most of the effect of precession.

Normally, *site data need not be extremely accurate*. Time accurate to five minutes, and latitude and longitude accurate to a degree or two, are generally good enough; much rougher settings are usually all right. (You can use the coordinates of a city up to 100 miles or 150 km away, maybe farther.) Errors in the site data have the same effect as a small error in leveling or positioning the mount, and once you align on two stars, the computer will eliminate them.

The exception is when you are relying on "one-star" altazimuth alignment. To locate the celestial sphere, the telescope must identify two points on it. The star is only one of those two points. The other is the zenith, whose right ascension and declination must be computed from the site data you've entered and the assumption that the mount is perfectly level.

In that situation, the site data should be accurate to 0.2° or better (1 minute of time) and the mount must be as level as possible. This is also the case if, when observing planets in the daytime, you must rely on "zero-star" approximate alignment. But it is usually much easier to align on two stars than to set up the tripod so accurately that one star will suffice.

### 3.3.2 Obtaining site data

Users of some telescopes can pick their country or state and city from a list; the rest of us have to type in our latitude and longitude, accurate to a degree or two. Latitudes and longitudes in Britain and North America can be estimated from the maps in Figures 3.3 and 3.4. For more accurate data, consult a large-scale map or atlas, look up your location at a geographical website such as http://www.gazetteer.de, or use a Global Positioning System (GPS) receiver. Some telescopes have a GPS receiver built-in (p. 29).

> *Hint:* Make sure you know whether you're using decimal degrees or degrees, minutes, and seconds. If your latitude is +45.75 and you enter it as 45°75′, the telescope will not accept it because the number of minutes is always less than 60.

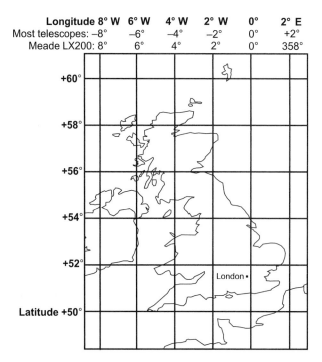

| Longitude | 8° W | 6° W | 4° W | 2° W | 0° | 2° E |
|---|---|---|---|---|---|---|
| Most telescopes: | −8° | −6° | −4° | −2° | 0° | +2° |
| Meade LX200: | 8° | 6° | 4° | 2° | 0° | 358° |

Figure 3.3. Latitude and longitude map of Britain.

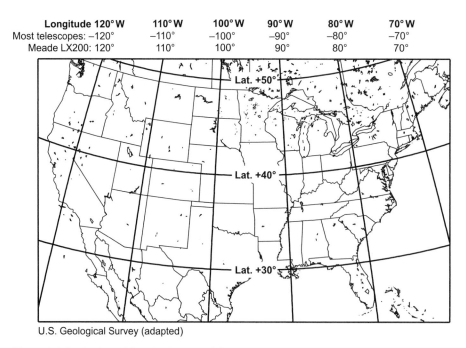

| Longitude | 120° W | 110° W | 100° W | 90° W | 80° W | 70° W |
|---|---|---|---|---|---|---|
| Most telescopes: | −120° | −110° | −100° | −90° | −80° | −70° |
| Meade LX200: | 120° | 110° | 100° | 90° | 80° | 70° |

U.S. Geological Survey (adapted)

Figure 3.4. Latitude and longitude map of the contiguous 48 states.

Instructions for setting the clock inside specific models of telescopes are given in Part II. These clocks, like all others, are imperfect; check their accuracy every month or two. Some of them use 24-hour format (e.g., 8 p.m. = 20:00); the rest require you to distinguish between a.m. and p.m. Clearly, if you mix up a.m. and p.m., the telescope will be totally mistaken about what is in the sky.

> *Hint:* Meade LX200 and Celestron NexStar telescopes require you to enter the date in American style – month/day/year. For example, 03/01/05 is 2005 March 1.

Meade LX200 telescopes reckon longitudes and time zone numbers in the opposite of the usual direction. Longitude $-84°$ (i.e., 84° west) and time zone $-5$ (Eastern Standard Time) go into an LX200 as $+084$ and $+05$ respectively. That's because the designers of the LX200 chose to measure longitude the same way on Earth as on other planets, i.e., increasing westward all the way around to $360°$.

If you wish, you can run your telescope on Universal Time (UT) regardless of your location; just enter the time zone number as 0. Be sure to give the date as well as the time in UT. For example, 8 p.m. EST on January 1 is 01:00 UT on January 2.

## 3.4 Why compasses don't point north

The setup procedures for all computerized telescopes assume that you know which way is north. In altazimuth mode, any error that you make will be corrected as soon as you align on a star. In equatorial mode, however, the polar axis must point exactly north for smooth, accurate tracking; we'll return to this in the next chapter.

The proper way to find true north is to sight on the star Polaris. Sometimes, however, you have to use a compass for initial orientation, and in parts of the world – especially Canada and the American West – compasses are surprisingly inaccurate.

The reason, of course, is that a compass points toward the north *magnetic* pole, not the pole of the Earth's axis. The two poles are not in the same place. What's worse, the north magnetic pole is a complex structure, not a single point, and moves around significantly from year to year.

The discrepancy between magnetic north and true north is called **magnetic deviation, magnetic declination**, or **compass correction** and is shown (for North America and Britain) in Figure 3.5. Clearly, if you try to find Polaris with a compass in Seattle, and you are not aware that your compass points 20° east of true north, your attempts to match the compass with the sky will be frustrating.

The good news is that *if you can find Polaris, you need not bother with any of this.* Polaris is always within 0.8° of true north.

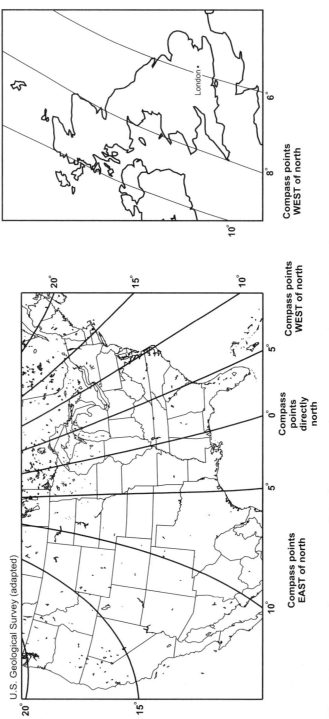

Figure 3.5. Magnetic deviation maps of North America and Britain. Data are for 1995 and can change 1° or more per decade.

## 3.5   Setting up the telescope

Detailed setup procedures for specific telescopes are given in Part II, but in general, here's how you set up a computerized telescope on an altazimuth mount:

(1)   *Level the tripod.*   Don't waste too much time on this, since exact leveling should not be necessary if you are going to align on two stars. I have personally found that Meade ETX-90 and Celestron NexStar 5 telescopes work well on tripod heads that are as much as 5° from level. But some NexStar 4 and 5 owners report that their telescopes *are* affected by errors in leveling and initial positioning, even after a two-star alignment, perhaps because of a firmware bug. If in doubt, do your own experiments.

Precise leveling is necessary in "one-star" and "zero-star" alignment modes. Also, leveling is important if you do not know the sky and are relying on the telescope to find its alignment stars without your help.

(2)   *Check the finder.*   Verify that the finder crosshairs indicate the center of the field of the main telescope; if not, adjust the finder by sighting distant land objects.

(3)   *Put the telescope in the "home" position*   as specified in its instructions (generally, pointing at the horizon, due south for LX200s, due north for NexStars and Autostars).

Again, great accuracy is not required unless you are performing a "zero-star" daytime alignment or are requiring the telescope to find the alignment stars accurately by itself. However, a large error may cause the telescope to report "bad alignment" because the computer thinks you have used the wrong stars.

(4)   *Choose an alignment mode.*   Depending on the telescope, there are generally several options:

- "Easy" or "auto" alignment, in which the telescope picks two stars and slews to their approximate positions, then waits for you to center each one;
- "Two-star" alignment on two stars that you choose;
- "One-star" alignment on one star that you choose;
- "Zero-star" alignment, in which the telescope relies entirely on the site data, tripod leveling, and precise home position.

"Easy" or "auto" alignment is usually best, but don't let it play tricks on you. The telescope doesn't go to the exact position of the alignment star, only the general area. (Accuracy depends on tripod leveling and placement in the initial home position.) A common pitfall is to center Castor ($\alpha$ Geminorum) when the telescope has chosen Pollux ($\beta$ Geminorum).

One-star alignment is rarely a good choice; it relies on perfect leveling of the tripod and accurate site data. Normally, if you can find one star, you can find two and get much better results.

Zero-star alignment exists only unofficially; manufacturers do not recommend it, but it is sometimes the best you can do. You can improvise a "zero-star" alignment by doing a one- or two-star alignment and telling the computer that each star is centered, without looking to see whether it really is. Zero-star alignment is sometimes useful for observing planets in the daytime but not much else. Expect pointing errors of several degrees until you can find bright stars or planets and sync on them. First-magnitude stars are visible with a medium-power telescope even in daylight.

(5) *Center each star in the field when told to do so,* then tell the computer that the star is centered. The more accurately you can center the star, the better, since small errors will be magnified as the telescope swings across the sky.

There – your telescope is set up and you can start observing. To confirm that it's working properly, start by going to one or two bright stars that you can identify.

> *Hint:* Always perform a "sanity check" on the alignment (or any other computer function) before relying on it. Tell it to go to one or two stars that you can easily identify, and make sure it does so. Make sure the telescope can find familiar objects before you ask it to find unfamiliar ones.

## 3.6 Choosing alignment stars

Of all the stars in the sky, which two should you align on? Actually, there are many pairs of stars that will work equally well, but here are some guidelines:

- The two stars should be on opposite sides of the zenith, at least 120° apart in azimuth.
- The two stars should not be at the same altitude; one should be appreciably higher than the other.
- Neither star should be within 20° of the zenith.
- One or both stars should be close to celestial objects that you are interested in observing. Pointing accuracy is best near the alignment stars.

Jim Burrows and Paul Rodman distribute (from http://www.ilangainc.com/bestpair) a computer program, *Best Pair II*, that calculates which alignment stars are best. It is based on a careful mathematical analysis of the computations that the telescope would perform on each pair. Normally, there is no single "best" pair at any given time; many different pairs of stars work equally well or almost equally well.

Do not use the Sun, the Moon, or a planet as an alignment star unless you have no other choice; positions of Solar System objects are not computed as accurately as those of stars.

## 3.7 Automatic setup with GPS

Some newer Celestron and Meade telescopes have a built-in GPS receiver for determining the latitude, longitude, and time, as well as a built-in magnetic compass and level sensor. Such a telescope can practically set itself up – or can it?

The answer is yes. Note however that you must still center the alignment stars yourself. (The obvious next step would be a photoelectric finderscope to do this automatically too!) What is automated is the initial process of entering the site data, leveling the telescope, and pointing it north. It takes a couple of minutes to acquire signals from GPS satellites; the built-in computer must sort through all the available signals, pick a set of them, and analyze them.

The main advantage of using GPS is to prevent blunders – you'll never make a big mistake entering the latitude, longitude, or time zone. As already noted, the extreme accuracy of GPS is not actually needed, and the GPS system does not improve the pointing accuracy of the telescope.

Moreover, a magnetic compass built into metal equipment is inherently inaccurate, and some telescopes (such as the initial NexStar GPS models) do not try to correct for magnetic deviation. As a result, the telescope can start out as much as $20°$ away from the first alignment star.

## 3.8 Tripods and piers

### 3.8.1 Steadiness

Really sturdy telescope mounts are not portable, and vice versa. At an observatory, you can tap the side of the telescope tube, and the image won't move; not so in the field. A good portable mount will shake for about one second after you lightly touch the side of the eyepiece; a light-duty mount will shake longer.

For visual astronomy, you can use a relatively light-duty mount as long as it doesn't shake while you're observing. Basic astrophotography requires a medium-duty mount, and serious astrophotography, a heavy-duty one. Based on personal experience, I classify the Meade ETX-90 mount as light-duty; the Celestron NexStar 5 as medium-duty; and the 8-inch (20-cm) LX90 at the low end of the heavy-duty range. The LX200 is sturdier than the LX90. Really heavy-duty mounts, some of them barely portable, are made by Losmandy, Astro-Physics, and Software Bisque (Paramount mounts).

There are two ways to control vibration: prevent it by using strong materials, or absorb it quickly after it starts. Unfortunately, these two strategies work against each other. The strongest material, steel, tends to "ring" for a long time when tapped. The best vibration-absorbers, wood and rubber, are subject to flexure. For permanent piers, steel embedded in concrete is much better than steel alone because concrete absorbs vibrations.

Figure 3.6. Vibration-reducing pads consist of a layer of soft rubber sandwiched between two layers of hard rubber.

Vibration in portable tripods can be reduced by putting rubber pads under the tripod legs; Celestron and Meade sell pads for this purpose (Figure 3.6). I find that they work miracles with Meade tripods (which are very hard and stiff) but don't help as much with Celestron tripods, which already have soft rubber feet for the same purpose.

A low tripod is a steady tripod, which is one reason many observers prefer to sit on a stool when observing. On the other hand, a tall tripod gets the telescope farther away from the ground and its associated air currents.

Occasionally, vibration is caused by the telescope's stepper motors. Such vibration is usually evident only at high powers and is very sensitive to the resonant frequency of the telescope and mount. This in turn makes it depend on the length of the tripod legs, the presence or absence of rubber pads, the weight of the eyepiece and dewcap, and the position of the telescope. With very lightweight mounts, the best tactic might be to turn off the drive when you want to use high power. On Meade Autostar telescopes, you can do this by selecting "Sleep telescope". If problems are persistent, relubrication of the worm gears may help.

### 3.8.2 Other tripod and wedge hints

It is important to tighten all the screws that hold a tripod and mount together; complex tripod heads have numerous bolts that can accidentally be left loose. In particular, make sure the central bolt that holds the LX90 or LX200 base to the tripod in altazimuth mode is good and tight.

Figure 3.7. A homemade template makes it much easier to center the LX200 on its tripod. This one is plywood, but cardboard would work just as well.

Do not overtighten the handscrews on tripod legs; you can easily apply enough force to crush the steel tubing. But do make sure they are secure. There are reports of one production run of Celestron heavy-duty tripods that were prone to collapse because a felt pad inside the leg would suddenly let go. The manufacturer has corrected the problem, but similar things can happen to other tripods; it pays to be alert.

The base of the Meade LX200 is hard to center on the tripod in altazimuth mode; you have to align a central bolt with a hole, both of which are hidden from view. A simple solution is to make a wooden or cardboard template that matches the base of the telescope and has a hole in the same position (Figure 3.7). Put the template on the tripod, then put the telescope on the template. A commercial device of this type, which also includes an accessory shelf, is the Scope Saver (Beigle/Bryant Engineering, P.O. Box 5247, Pine Mountain Club, California 93222, U.S.A., http://www.scopesaver.com).

### 3.8.3 Observatories and permanent piers

An observatory dome can be a godsend in a windy environment, but elsewhere, domes are often troubled by unsteady air flowing into or out of the slot. "Clamshell" domes that expose at least half the sky are better; so are buildings with removable roofs. The largest observatories use domes with slots because nothing else can be built big enough.

A permanent pier made of steel is likely to "ring" when tapped, continuing to vibrate for several seconds. Steel embedded in concrete, or at least filled with sand, is much better than steel alone. The pier should have a diameter comparable to the aperture of the telescope, if not larger. Mine, made of 4-inch (110-mm) steel pipe, is definitely shaky with an 8-inch (20-cm) telescope; I bolted heavy pieces of wood to the sides of it to stiffen it and hope one day to surround it with a concrete cylinder.

## 3.9 Pointing accuracy

### 3.9.1 What to expect

The pointing accuracy of a low-cost computerized telescope is necessarily limited. Affordable amateur telescopes do not get objects perfectly centered in the field every time. In general, a portable computerized mount is performing acceptably if it gets objects within the field of a 26-mm eyepiece. This corresponds to a maximum error of about 20' (0.33°) with an 8-inch (20-cm) $f/10$ telescope.

> *Hint:* Do not judge a telescope by how well it finds the two alignment stars at the start of the alignment procedure. At that point, it's just guessing, and its accuracy depends on tripod leveling, site data, and accurate placement in the home position – factors that will not affect it after aligning on two stars.

Pointing error can be measured three ways. **Maximum error** is, of course, the greatest error encountered in a series of tests. **Average error** is the average (mean) of the errors, treating all of them as positive numbers (because, of course, opposite errors in successive tests do not cancel out). **RMS (root-mean-square) error** is the square root of the mean of the squares of the errors, often used by professional astronomers because it gives greater weight to large errors.

### 3.9.2 Factors that affect pointing accuracy

Despite what people think, precise site data and accurate placement in home position are not necessary for accurate pointing, at least as long as you align on two stars. (One-star alignment is inherently dubious, and zero-star alignment is always bad.) The factors that affect pointing accuracy are:

(1)     *Choice of alignment stars.*   See p. 28. If you use two alignment stars that are too close together, any small error in the position of one star will be magnified as it is extended around the sky.

(2)     *Accuracy of initial alignment.*   It is important to *center* all alignment stars (including stars that you sync on, later in the session), not just get them into the field. An eyepiece with crosshairs is helpful. Do not sync on the Moon or a planet, since their positions are not calculated accurately enough. Do not change focus

between one alignment star and the next, since changing focus will shift the image slightly.

Before you sync, *wait* and make sure the telescope has finished slewing; the ETX-90 slews, almost silently, for as much as 45 seconds after the audible motion has stopped, and this is typical. But do not delay unduly between one alignment star and the next; telescopes generally assume that you will find the two stars in rapid succession.

(3) *Tripod leveling*   in altazimuth mode only, and only when aligned on just one star. The cure is to align on two stars. (Some telescopes do seem to be affected by leveling and by accuracy of placement in the home position even after a two-star alignment.)

(4) *Polar axis alignment*   in equatorial mode only, when aligned on one star, or one star and Polaris; see the next chapter for details.

(5) *Backlash (slack in the gears).*   On the NexStar 5, always make your final approach to each alignment star by moving upward and to the right, to take up slack. Similar tactics may help with other telescopes.

On Meade Autostar telescopes, be sure to run the "Train Drive" procedure described in the manual; the computer will then store the compensation factors. Pointing accuracy will be poor until you do this. Since the gears smooth out with use, repeat the procedure after the first few sessions and again every few months.

(6) *Flexure and mirror shift.*   If any part of the tripod or mount bends as the load shifts its position, pointing accuracy will be affected. The mirror of a Schmidt–Cassegrain or Maksutov–Cassegrain telescope shifts slightly as the telescope moves from one part of the sky to another; this is inevitable because the mirror has to be movable for focusing.

(7) *Balance.*   Fork-mounted telescopes should not be perfectly balanced; some imbalance helps take up slack in the gears. But if the direction of the imbalance changes, accuracy will suffer.

For instance, the NexStar 5 is designed to be a bit front-heavy. A heavy eyepiece can shift the balance and impair pointing accuracy. Adding a dew cap on the front of the telescope is a quick cure.

(8) *Encoder and gear accuracy.*   There is little you can do about nonlinearities in the encoders and the gears that drive them, except to be aware that gears become smoother with use. If, on one particular occasion, pointing accuracy is not very good, the best thing to do is often to realign. You may have hit a bad spot on the gears the first time.

(9)   *Nonperpendicularity.*   Two-star alignment assumes that the telescope tube is perfectly perpendicular to the declination (or altitude) axis, which in turn is perfectly perpendicular to the polar (or azimuth) axis. That is why you are advised never to remove the telescope from the fork mount. It's also the main reason why some telescopes work better than others of the same make and model.

### 3.9.3 Aligning the telescope tube in the mount

If your telescope is out of warranty, you may be able to adjust the fork mount to improve perpendicularity. For instance, the Meade LX200 tube is held to the fork by screws that are in slots so the tube can be lined up. This procedure is called *collimating the mount*, not to be confused with collimating the optics. Do not attempt it on telescopes that are still under warranty.

You can also adjust the declination setting circle so that it reads 90° when the telescope is pointed directly away from its base. This is not a risky operation and does not void the warranty.

The hard part is finding out when the tube is in the right position. What you want to do is point the tube directly away from the base and get it perfectly parallel to the polar axis.

One way to do this is to aim the telescope straight up at the starry sky. Then swivel the telescope around its polar axis and see whether the same point remains in the center of the field. To get it to do so, you will need to slew in declination as well as adjusting the position of the tube. When you've achieved the right position, the declination setting circle (if any) should read 90°; if it doesn't, adjust it.

Another method, a bit less precise, is to level the base of the telescope with a bubble level, then confirm that the front of the tube is also level.

A very precise method, popular with Schmidt–Cassegrains, is to reflect a laser pointer off the corrector plate at any convenient angle, then adjust until the reflection remains stationary as the telescope swivels. Schmidt–Cassegrain corrector plates are not flat, but they are circularly symmetric, and that is enough. The same technique may work with Maksutov–Cassegrains and refractors.

### 3.9.4 The double-GO TO trick

Sometimes, if you tell the telescope to go to the same object twice in succession, it will be considerably more accurate the second time. I have noticed this particularly with a NexStar 5. Choose the object, press ENTER (or GO TO if it's a Meade telescope), let the telescope finish slewing, and then press ENTER or GO TO again.

A second GO TO is also a good way to compensate for poor tracking. Low-end computerized telescopes can find objects a good bit more accurately than they can track them because the whole calculation is not updated continuously. Thus a second GO TO will often retrieve an object that has drifted out of the field. On Autostar telescopes, a double GO TO initiates a square-spiral search (p. 206).

### 3.9.5 Meade high-precision mode

To get maximum accuracy with any computerized telescope, simply sync on a known star in the same part of the sky before going to the faint object that you want to see.

High precision mode on the Meade LX200 and Autostar automates this process. The telescope chooses an appropriate star and asks you to sync on it before going to the object you have chosen.

In high-precision mode, the LX200 generally achieves an RMS error of 1', sufficient for pointing CCD cameras at visually invisible deep-sky objects. However, after syncing on a star near Polaris or the zenith, pointing accuracy elsewhere in the sky tends to suffer. The cure is to sync on a star high in the southeast or southwest.

### 3.9.6 *TPoint* Software

By aligning on more than two stars, the computer can measure non-perpendicularity, flexure, and other sources of error, then correct for them. This is the purpose of *TPoint*, a product of TPoint Software (http://www.tpsoft.demon.co.uk, Abingdon, Oxfordshire, England). *TPoint* is used at large observatories but is also marketed to amateurs by Software Bisque (http://www.bisque.com), integrated with their *TheSky* online star atlas. It runs on a personal computer (normally a laptop) connected to the telescope.

With a portable amateur telescope, it is normal to align *TPoint* on 10 to 50 stars (the more the better) and achieve a pointing accuracy better than 2'. (Much higher accuracy is not possible because of mirror shift and other nonrepeatable errors.) In equatorial mode, *TPoint* also tells you how accurate your polar alignment is and what to do to correct it.

Permanently mounted telescopes can be aligned on dozens of stars for extremely high precision. With the professional version of *TPoint*, the 4-meter Anglo-Australian Telescope (in Siding Spring, N.S.W.) achieves a pointing accuracy of one arc-second.

### 3.9.7 What the telescope is calculating

The relation between altitude, azimuth, right ascension, and declination is:

$$\sin a = \sin \delta \sin \phi + \cos \delta \cos \phi \cos(S - \alpha)$$

$$\cos A = \frac{\sin \delta - \sin \phi \sin a}{\cos \phi \cos a}$$

where:

> $a$ = altitude (zenith = 90°)
> $A$ = azimuth (north = 0°, east = 90°, up to 360°)
> $S$ = sidereal time, expressed as an angle (24$^h$ = 360°)
>  = right ascension of the zenith
> $\alpha$ = right ascension of object, expressed as an angle
> $\delta$ = declination of object
> $\phi$ = latitude of observing site, positive if north
>  = declination of zenith

By aligning on two stars with known $\alpha$ and $\delta$, the computer can solve for $\phi$ and $S$, then keep track of the passage of sidereal time. If the base is not level, the calculation is similar, but $a$, $A$, $S$, and $\phi$ are referred to the telescope's main axis rather than the true zenith. Equatorial-mode two-star alignment is merely a special case of a nonlevel base.

Near the zenith and near Polaris, the equations are harder to solve for two reasons. In those parts of the sky, a small change in position corresponds to a big change in azimuth or right ascension respectively. Also, when $\sin a \approx 1$ or $\sin \delta \approx 1$, accurate results require arithmetic with a large number of decimal places; a small change in the sine reflects a large change in the angle. That is why you are advised to avoid alignment stars in those areas.

The actual algorithms used in telescopes are proprietary. NexStar telescopes reportedly work well with Polaris as an alignment star; Meade telescopes reportedly do not.

To learn more about calculations such as these, see *Practical Astronomy With Your Calculator*, by Peter Duffett-Smith (Cambridge, 1989), and *Telescope Control*, by Mark Trueblood and Russell M. Genet (Willmann-Bell, 1997).

## 3.10 Computer control

You can connect your telescope to an external computer to obtain better control functions and a richer user interface. The usual method is to align the telescope in the usual way, then hand over control to the computer, which is connected through its serial port. The same computer may also be running a CCD camera, autoguider, and/or observation logging software.

All of the major computerized star atlas packages can control telescopes. These include *Starry Night* (http://www.starrynight.com), *TheSky* (Software Bisque, http://www.bisque.com), and *SkyMap* (http://www.skymap.com), among others. The on-screen star map shows where the telescope is pointed; click on an object on the screen, and the telescope slews to it automatically. (See also p. 35 for information about *TPoint*, a software package that measures and corrects the errors in your telescope mount, giving extremely high pointing accuracy.) Because software is evolving so rapidly, it would be premature for me to recommend a specific product; instead, you should download the trial versions and see them for yourself.

Using a laptop computer in the field raises some practical concerns. The screen is generally bright enough to interfere with night vision. Switching the software to "night vision mode" (all-red display) reduces but does not eliminate the problem. It is better to leave the display in its normal configuration and put a piece of red plastic in front of it. Deep red transparent acrylic sheets are sometimes available at glass shops or sign shops. "Gels" (cheap, flexible filters for theatrical lights) are available from photographic and theatrical suppliers; use the darkest shade of red available.

The laptop should have an electrically isolated power supply. This will be the case if it is operating off its own batteries, the AC power line, or an inverter. Electrical noise may be a problem if the laptop is powered directly from the same battery as the telescope, because then there will be a *ground loop* (a pair of alternative DC paths between the same two pieces of equipment). Note that with Meade LX200 telescopes, the frame of the telescope is not (and should never be) connected to the negative side of the battery.

But you don't actually need a laptop – a palmtop will do. At least one program (*TheSky*) is available for palmtop computers that are no bigger than the telescope's original control box. Palm computers may soon supersede control boxes entirely.

For remote control, serial cables can be hundreds of meters long without suffering appreciable signal degradation. The same is not true of the USB, SCSI, or parallel-port cables that interface to CCD cameras.

## 3.11  Electricity for telescopes

Most if not all computerized telescopes run on 12 volts DC. Several will also run on lower voltages. The Meade LX200 nominally requires 18 volts, but in fact my 8-inch (20-cm) runs quite contentedly on 12 volts; all that is lost is some slewing speed.

In the field, rechargeable lead-acid or NiMH batteries are convenient power sources. The capacity of a battery is specified in ampere-hours:

$$\text{Battery life (hours)} = \frac{\text{Capacity (ampere-hours)}}{\text{Load (amperes)}}$$

A typical telescope draws about 0.5 ampere, or a bit more, and will therefore run all night from a 7-ampere-hour battery, but CCD cameras, dew removers, and other accessories add to the load. I use a 17-AH battery pack designed for jump-starting automobiles; I've removed its huge cables and added some small, convenient sockets. Car batteries are not designed for deep discharge and are not suitable.

The voltage of a battery is nearly constant until the battery is almost exhausted. For example, a 12-volt lead-acid battery may charge up to 13.1 volts, then settle at 12.3 to 12.6 V during most of the discharge cycle. When it's nearly

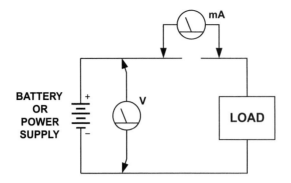

Figure 3.8. Voltage can be measured across the battery at any time, but current (amperage) can only be measured by interrupting the circuit and inserting the meter into it.

empty, it will drop below 12 V just before giving up the ghost. A digital voltmeter is useful for monitoring these small changes in voltage.

Figure 3.8 shows how basic electrical measurements are made. Voltage is the pressure that drives a current through a resistance. Thus, to measure current (amperage) you must break the circuit and insert the meter into it, but voltage is measured across the battery.

There are only three ways to change a voltage: throw away some of the energy with a resistance (as is done inside the LX200 hand box, which is why it gets warm); use a transformer; or use a switching regulator. Transformers trade volts for amperes, but they work only with alternating current. Switching regulators, such as Meade's 12-to-18-volt DC converter, do much the same thing. They chop the incoming DC into high-frequency AC; run it through a transformer or an inductor that provides transformer-like action; and then rectify it back to DC, meanwhile regulating the voltage.

Current is measured in amperes (amps, A) or milliamperes (mA), where 1000 mA = 1 A. The current rating of a power supply is the *maximum* it can deliver, while the voltage rating is the *actual* value; thus you can connect a 12-volt, 2-ampere power supply to a 12-volt, 0.5-ampere telescope and only 0.5 ampere will flow.

Wattage (power) is a measure of the rate at which energy is being expended. It is the product of voltage and current:

Watts = Volts × Amperes

Thus a 12-volt, 0.5-ampere dew heater consumes 6 watts.

# Chapter 4
# Using equatorial mounts and wedges

## 4.1 Why equatorial?

Table 4.1 sums up the advantages and disadvantages of equatorial and altazimuth mounts.

There are two main reasons for using an equatorial mount (such as the one in Figure 4.1): to eliminate field rotation (Figure 4.2) in long-exposure photography, and to establish which way is north in the sky so that you can use charts and measure double-star position angles. Apart from that, altazimuth mode is almost always preferable. Setup is simpler and quicker, and you don't need an **equatorial wedge** to tilt the base.

There is one situation in which an equatorial mount is easier to set up than an altazimuth one. That is when you have a permanent telescope stand that is accurately aligned with the Earth's axis. In that case, all you have to do is attach the telescope and sync on one star. That's all – the telescope is aligned, calibrated, and ready for both visual observing and photography.

## 4.2 Must field rotation be eliminated?

Equatorial mounts get rid of the field rotation illustrated in Figure 4.2. Celestial objects tilt as they rise, travel across the sky, and set. An equatorially mounted telescope tilts with them, so that everything remains stationary in the field of view, but altazimuth-mounted telescopes suffer field rotation. With an altazimuth mount, the object that you're tracking remains centered, but everything else rotates around it.

Can you make do with an altazimuth mount? Maybe. Field rotation does not affect photographic exposures shorter than about 15 seconds, so it is no obstacle to solar, lunar, or planetary imaging; only deep-sky photography is affected, and even then, altazimuth-mode exposures of several minutes are possible when imaging an object low in the eastern or western sky. **Field de-rotators** are available for some of the larger altazimuth-mounted telescopes; they rotate

Table 4.1 *Altazimuth versus equatorial mounts*

*Altazimuth mounts*
Simple and lightweight (no wedge)
Stable (centered load, low center of gravity)
Easy to set up (point the axis roughly straight up,
    then sight on two stars)
Tracks using two motors
Can see entire sky
Hard to maneuver when pointing near zenith
    (some models cannot aim straight up
    with a camera attached)
Easy to maneuver when pointing near Polaris
Unsuitable for long-exposure photography (Figure 4.2);
    OK for photographing Sun, Moon, and planets

*Equatorial mounts*
Heavier and more complex
Must be aligned on Earth's axis
    (by sighting Polaris and/or checking for drift)
Smoother tracking (just one motor running)
May not be able to see far southern sky (Figure 4.14)
Easy to maneuver when pointing near zenith
    (also easier to reach Meade ETX focus knob,
    which is not hidden between fork arms)
Hard to maneuver when pointing near Polaris
    (some models will not aim near Polaris
    with a camera attached)
Easy to slew directly north, south, east, or west
    for direct comparison to star map
Required for long-exposure photography

the camera to match the stars but are usable only when photographing through the telescope, not when using telephoto lenses or other instruments mounted "piggyback" on it.

Even equatorial mounts suffer from field rotation if they are several degrees away from correct polar alignment. This is the subject of many misconceptions. For detailed analysis, see *Astrophotography for the Amateur* (Cambridge, 1999), Appendix A, and the software at http://www.covingtoninnovations.com/astro. Here are some basic rules of thumb:

(1)     The field always rotates around the star that the telescope is tracking (the guide star, Figure 4.3). If the camera is aimed in a different direction than the telescope, field

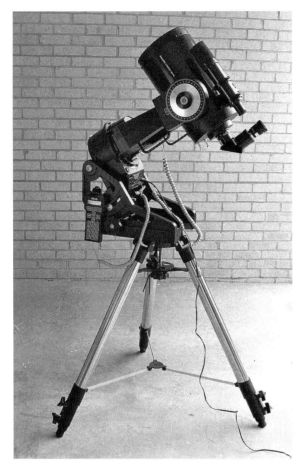

Figure 4.1. Meade LX200 telescope equatorially mounted on Meade Superwedge and field tripod. Compare Figure 3.1.

rotation will turn stars into arcs of circles centered on the guide star, which lies outside the picture.

(2) Many lenses have an aberration that looks just like field rotation but is always centered on the center of the field and is the same regardless of exposure time. Do not be misled by this. Such an aberration usually disappears when the lens is stopped down.

(3) A picture will normally look sharp if there is no more than $0.1°$ of field rotation, provided the guide star is near the center of the picture. Curiously, this is true regardless of the focal length or magnification, because telescopes don't magnify rotation. (A rotating wheel, viewed through a telescope, still rotates at the same speed.) Lunar and planetary work can tolerate somewhat more field rotation because the highly magnified images are not as sharp.

(4) With a telescope in altazimuth mode at latitude $40°$, the rate of field rotation varies from $0.5°$/minute along much of the meridian to just $0.07°$/minute low in the east

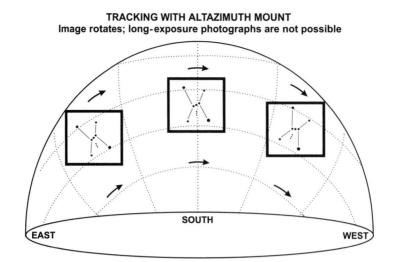

**TRACKING WITH ALTAZIMUTH MOUNT**
**Image rotates; long-exposure photographs are not possible**

SOUTH

EAST                    WEST

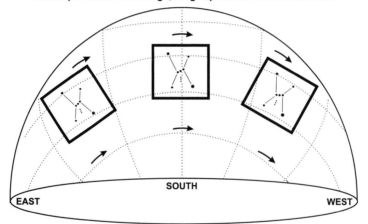

**TRACKING WITH EQUATORIAL MOUNT**
**Telescope rotates with image; long exposures work as intended**

SOUTH

EAST                    WEST

Figure 4.2. Field rotation. An equatorial mount tracks the changing tilt of celestial objects so that long-exposure photographs are possible.

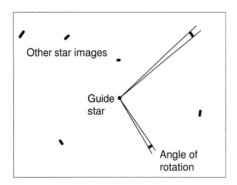

Other star images

Guide
star

Angle of
rotation

Figure 4.3. Effect of field rotation on a photograph. The rotation is always centered on the star that the telescope is tracking. (From *Astrophotography for the Amateur*, Cambridge, 1999.)

or west. Thus, exposures of one or two minutes are practical in some parts of the sky.

(5) A perfectly polar-aligned telescope would have no field rotation at all. For a maximum of 0.1° rotation in a 1-hour exposure of objects high in the sky, the telescope must be polar-aligned to within 0.4°. That level of accuracy is easy to achieve with one or two iterations of the drift method (p. 49). Much of the time, simply sighting on Polaris is good enough.

## 4.3 Using an equatorial mount

### 4.3.1 Setting up the mount

Setting up an equatorial mount requires the mount itself – that is, the wedge on top of the tripod – to be aligned with the Earth's axis. No subsequent adjustment of the telescope or computer can correct an error in polar alignment.

You have, therefore, three tasks: level and adjust the tripod, point the wedge so that the low side of it is due north,[1] and adjust the wedge by sighting on Polaris.

Most wedges have adjustments for latitude (polar altitude) and azimuth, but it's quite possible to do without them. My best tripod is made of solid oak and is cut for a latitude of 34° north; I make small adjustments by moving the legs in and out.

The purpose of leveling is to help you get a previously adjusted mount back into the same position. The idea is that if you level the tripod before adjusting the wedge, then the next time you set it up, you can just level the tripod and the wedge will again be at the correct inclination. Intrinsically, an equatorial mount doesn't care whether the tripod head, or anything else below the wedge, is level; all that matters is that the wedge is perpendicular to the Earth's axis.

### 4.3.2 Rough polar-axis alignment without sighting stars

If you can see Polaris in the sky, you can skip this section and go directly to the next one. If, however, you are setting up your mount in daylight, proceed as follows to get it roughly aligned.

Find true north by using a compass and correcting for magnetic declination (p. 26). Orient the mount so that the polar axis points north. Adjust the tilt of the wedge ("latitude" or "altitude") to match your latitude (Figure 4.4). For example, in New York, at latitude 40°, the baseplate should be 40° from vertical.

---

[1] This entire book, and especially this chapter, is written for observers in the Northern Hemisphere. South of the equator, everything works in the opposite direction, and instead of Polaris, you have to sight on the much fainter star σ Octantis.

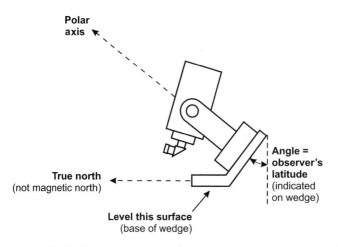

Figure 4.4. Setting up an equatorial mount when you cannot see Polaris.

With larger wedges, this is easy because latitudes are marked on the wedge itself, and there is a bubble level on the base. If the only bubble level is on the telescope (as with the Meade ETX-90), you can level the tripod in altazimuth mode, then incline the base to the appropriate angle (Figure 4.5).

Then refine the alignment by sighting on Polaris and/or by using the drift method (p. 49). Note that in the daytime, you can use the drift method on a bright star; stars down to second or third magnitude are visible in the telescope at high magnification. You can also use the drift method on planets, but not the Moon, whose orbital motion is fast enough to interfere.

### 4.3.3 Finding Polaris

Alone of all the stars, Polaris is always in the same position in the sky. Ursa Major and/or Cassiopeia always indicate its position. There are no other bright stars near it.

Figure 4.6 shows the northern sky; put the current date at the top to see how it looks at 9 p.m. local time. At latitudes south of New York, part of the bottom of the diagram will be very close to the horizon, probably concealed by trees. Farther north, the whole circle will be high in the sky and a patch of sky will be visible below it.

The key to finding Polaris is to find Ursa Major or Cassiopeia, or both if possible, and let them point the way. Note that Polaris is the only bright star in its neighborhood. Note also that Ursa Minor (the Little Dipper, containing Polaris) is not prominent and does not look like a dipper.

In a 4-inch (10-cm) or larger telescope, you can confirm your identification of Polaris by seeing its 9th-magnitude companion 180″ away.

The only other star you are likely to mistake for Polaris is Kochab (β Ursae Minoris). It is about the same brightness as Polaris but is appreciably yellower and has no close companion.

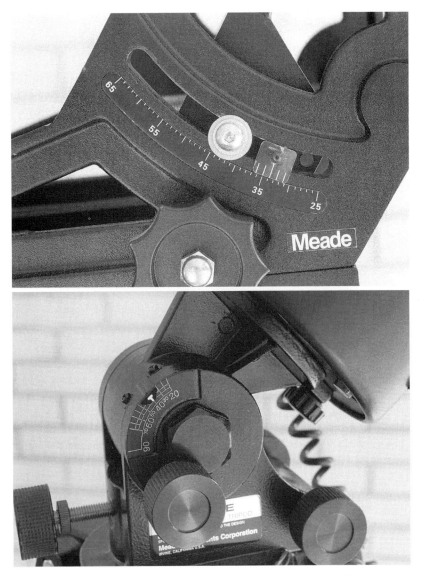

Figure 4.5. Scales on the wedge or tripod indicate correct inclination for your latitude. Meade Superwedge (top); Meade ETX-90 field tripod (bottom).

### 4.3.4 Rough polar-axis alignment on Polaris

Your task is to find Polaris, then aim the telescope directly away from its base (Figure 4.7) and center Polaris by adjusting the tripod. The computer in the telescope is not involved, and if you can unlock the brakes and move the telescope manually, the computer need not even be turned on.

After centering Polaris, you can obtain greater accuracy by moving about 0.8° toward Cassiopeia, away from Ursa Major.

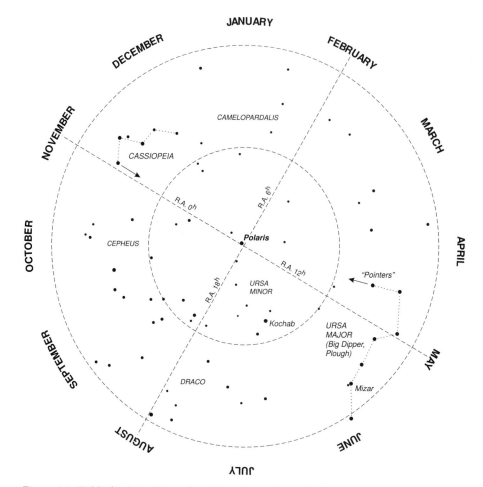

Figure 4.6. Field of Polaris. To see the orientation at 9 p.m. local time, put the current date at the top.

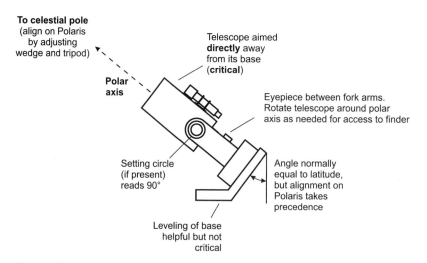

Figure 4.7. During polar alignment, the telescope points directly away from its base.

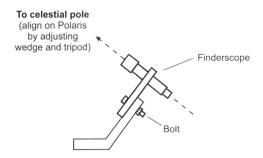

**To celestial pole**
(align on Polaris
by adjusting
wedge and tripod)

Finderscope

Bolt

Figure 4.8. A simple tool for aligning a wedge before putting the telescope on it.

Actually doing this may not be easy. One challenge is getting the telescope pointed directly away from its base. Meade LX200 and ETX telescopes have setting circles to tell you when they are in this position; the NexStar 5 does not. Even if you have setting circles, they may not be accurate (see p. 34). The computer is no help because it can't read declinations accurately until you have sighted on some stars. Don't panic; if you point the telescope away from its base "by eye" you'll get close enough to start. Later, you can find the proper position with a bubble level and mark it.

A more serious problem is that you may not be able to look through the telescope or its finder while it is in this position. You are free to swivel the telescope around its polar axis (slew in right ascension) as needed, since the telescope is aimed directly away from its base the whole time. But even so, the eyepiece and/or finder may remain inaccessible between the fork arms.

In that case, the gadget in Figure 4.8 will prove indispensable. It enables you to line up the mount before putting the telescope on it. I made my own, with an old finder and a block of wood, but commercial versions exist. Drilling must be done carefully, with a drill press, for good perpendicularity.

Remember that *Polaris is 0.8° away from the true pole, in the direction of the fainter end of Cassiopeia, away from the handle of the Big Dipper.* As far as possible, line up on that point rather than on Polaris itself.

## 4.3.5 Initializing the computer

Once you've set up the mount, it's time to turn on the computer and align the encoders on one or two stars.

Make sure the computer is in equatorial ("polar") mode, of course, or great confusion will ensue.

### One-star alignment
On a perfectly polar-aligned mount, one star is sufficient to locate the whole celestial sphere, because the telescope can assume that its axes correspond directly to right ascension and declination.

This is how the Meade LX200 (non-GPS) handles alignment in equatorial mode. All you really have to do is turn the telescope on, identify one star, and sync on it; the telescope does the rest. You do not have to choose "Align" on the keypad.

If you go through the alignment procedure on the Meade LX200 or Autostar keypad, what you get is a bit fancier. You start by putting the telescope in the specified home position: declination $+90°$, tube upside down (LX200) or right side up (Autostar). The telescope then slews to the estimated position of Polaris (slightly off the true pole) to help you adjust the mount, and then chooses an alignment star and moves to it for you to sync (see below). This is basically a one-star process because the position of Polaris is estimated, not measured.

With one-star alignment, pointing accuracy is completely at the mercy of polar axis alignment. This isn't as bad as it sounds because you can use pointing accuracy as a test of polar alignment. After initializing, go to stars in several parts of the sky and make sure they are all found successfully. If not, refine the polar axis alignment.

### Two-star alignment

The Celestron NexStar takes a radically different approach. It treats an equatorial mount as a tilted altazimuth mount, and just as in altazimuth mode, it initializes on two stars (other than Polaris). This enables it to find objects accurately even if the polar axis is misaligned, but it also means you can't use pointing errors to detect polar axis error.

You can still refine the alignment by the drift method (p. 49) provided the tracking mode is set to equatorial. Otherwise the telescope will track in two dimensions, just as it does on an altazimuth mount, and the drift method won't tell you anything.

Meade Autostar telescopes, including the LX200 GPS, offer both one-star and two-star alignment. Two-star alignment gives much better pointing accuracy and I strongly recommend it.

## 4.4 Refining the polar alignment

### 4.4.1 Iterating on Polaris and one other star

When a Meade telescope is initialized on Polaris and one other star, it assumes that the date, time, and location have been set accurately and that the tube was placed accurately in the initial home position specified for that type of telescope.

You can generally improve the polar alignment by repeating this procedure once or twice, because each time, the telescope will compute the apparent position of Polaris more accurately.

Contrary to the LX200 manual, you do not need to wait 15 minutes between iterations. Nor do you need to step through the alignment process on the menus. It is sufficient to go to Polaris, adjust the mount, and then go to a star and center and sync on it, over and over.

However, other precautions are in order. For your alignment star, don't use Hamal (α Arietis), Arcturus, or Spica. Why? Because Hamal has about the same right ascension as Polaris ($2^h30^m$), so when comparing Hamal to Polaris, the computer can't distinguish polar alignment error from declination error. Arcturus and Spica, near $14^h30^m$, are $180°$ away from Polaris, and if you use them, you'll find yourself bouncing back and forth between two equal and opposite polar alignment errors.

In any case, it is generally better to correct only *half* the error each time you adjust the mount, since otherwise, the telescope will often overshoot the desired position.

There are other obscure conditions under which this technique does not work well; in particular, I have never had much luck with it on the ETX-90. If iteration doesn't seem to be helping, don't waste much time with it, but instead, go directly to the drift method (next section).

### 4.4.2 Fine alignment – the drift method

Whether or not you can see Polaris, you can achieve extremely accurate polar alignment by checking whether stars drift north or south during tracking. For this purpose, use a high-power eyepiece, preferably one with crosshairs.

Start by tracking a star that is high in the southern sky. If your polar alignment is good, the star will remain centered in the field for five minutes or more. Ignore any periodic east–west shifting, which is caused by irregularities in the gears, and look only for a steady movement north or south.

Note the direction and amount of drift over a specific period, such as one or two minutes. Don't go too long, or you may be misled by flexure or mirror shift.

Then try to correct the drift by moving the polar axis slightly to the left or right. One try will tell you whether you've lessened the problem, made it worse, or perhaps even reversed it by going too far. Then try again. Repeat until the drift is practically zero.

You'll find that if the star drifts *north*, the polar axis needs to move to the *right* (east), and if it drifts *south*, the polar axis needs to move *left* (west). But if – like me – you find this rule hard to memorize, don't worry. The first time you move the polar axis, you'll find out whether you moved it the right way.

Next, track a star that is about halfway up in the east or west (at least $15°$ above the horizon). This will tell you whether the inclination (tilt) of the polar axis is correct. If a star in the *east* appears to drift *north*, the polar axis needs to move *down*, and if it drifts *south*, the polar axis needs to move *up*. Use the altitude ("latitude") adjustment on the wedge or pull the north or south tripod leg in or out.

## 4.5 Using wedges

To attach a larger Schmidt–Cassegrain telescope to a wedge, you must bolt it to three holes. The manufacturers expect you to insert three screws and tighten

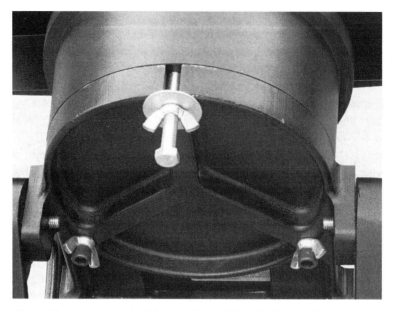

Figure 4.9. Author's method for attaching a Schmidt–Cassegrain telescope to a wedge. Long bolt (at top) is put in place in advance, wing nut is tightened with one hand while wedge supports weight of telescope. Then thumbscrews are placed in the two remaining holes.

them with hex wrenches every time you do this. Finding this procedure tedious and clumsy, I've simplified it in two ways (Figure 4.9).

First, one long bolt, with a wing nut and washer on it, is put in place before I pick up the telescope. Then I place the telescope onto the wedge, sliding this bolt into the slot, and tighten the wing nut with one hand while holding the telescope with the other.

The telescope is then secure, and, without tools, I can install thumbscrews (Figure 4.10) in the two remaining holes. These thumbscrews are homemade, built by joining wing nuts to bolts or pieces of threaded rod with high-strength epoxy. For strength, the wing nut is tightened against the bolt head or another nut, so that it would be moderately strong even without the epoxy.

*Be sure not to insert screws too deeply into the telescope base*, as they can bump into the internal mechanism and damage it (see p. 171). The manufacturer's original screws are a good indication of how deep you can safely go. Little or no strength is gained by threading a screw farther into a hole than a distance equal to its own diameter, and nothing at all is gained by going past the threaded portion of the hole.

Also, do not overtighten screws. A 9-mm screw can easily support a ton – but much or all of this ton may consist of the tension with which the screw is tightened! Screws that carry heavy weight should be tight enough to be snug, but no more. Excessively vigorous use of a wrench can break screws and strip threads.

Beware of cross-threading. Telescope mount castings, tripod heads, and wedges are generally aluminum; screws are steel. If the two come into conflict,

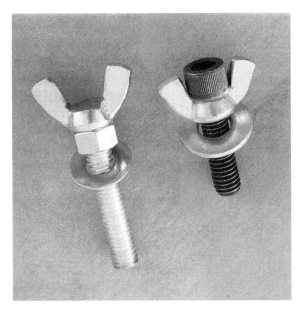

Figure 4.10. Homemade thumbscrews. Left: wing nut, ordinary nut, and threaded rod.
Right: wing nut and bolt. High-strength epoxy holds the wing nuts in place.

steel will win. You can easily ruin the threads in a hole by putting a steel screw
into it at an angle.

Use a sturdy wedge. The Meade Superwedge, designed for the 12-inch LX200,
is usable with the 7-inch, 8-inch, and 10-inch. It is very steady but somewhat
hard to adjust in altitude and azimuth. Alternatives to the Superwedge are
marketed by Ken Milburn (Bonney Lake Astro Works, 20508 125th Street Court
East, Sumner, WA 98390, U.S.A.) and by Jim Mettler (3200 West 450 North,
West Lafayette, IN 47906, U.S.A.). Both are elegantly machined, shiny, smooth-
operating, and not a great deal more expensive than the Superwedge.

If you are content to make adjustments by moving the tripod legs, or if the
wedge is to be installed on a permanent pier, it is easy to make your own wedge,
cut for a fixed angle matching your latitude. Wood is a good material since it
absorbs vibration. Figure 4.11 shows the wooden wedge on my permanent pier.

Adjustable wedges are harder to make, at least for heavy telescopes; one
design that works well with lighter telescopes uses a hinged panel and a turnbuckle, relying on the telescope's off-center weight to keep the turnbuckle tight.

## 4.6 Tracking in equatorial mode

### 4.6.1 Tracking rates

Diurnal motion does not affect all celestial objects equally, but the difference is
so small that you can usually ignore it. Because the Earth orbits the Sun, the

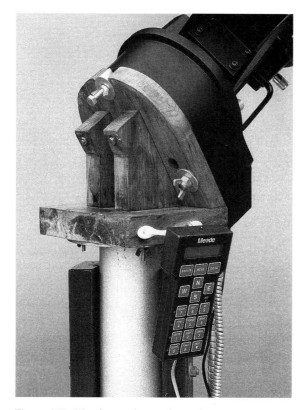

Figure 4.11. Wooden wedge on the author's permanent pier, made from cedar by the late Elmo Mauldin, woodworker.

stars seem to gain four minutes per day relative to solar time; that is, a **sidereal day**, measured between successive meridian passages of the same star, is only 23 hours and 56 minutes.

Accordingly, telescope drives offer a choice between **solar rate** (one rotation per 24 hours) and **sidereal rate**, which is about 2.7% faster.

The difference between sidereal and solar rate is hardly noticeable. If you track the stars at the solar rate, the total tracking error will be about one degree per day, or 0.1° in two and a half hours.

Celestron telescopes also offer **King rate**, defined by E. S. King at Harvard in 1931, which is slightly faster than sidereal rate in order to compensate for atmospheric refraction.

**Lunar rate** is more problematic. The Moon's orbital motion is fast enough to make sidereal and solar rates unsuitable; if you track the Moon at the sidereal or solar rate, it will move completely out of the telescope field in an hour or so. Ignoring parallax, lunar rate should be 3.3% slower than solar rate, and that is what most telescope drives deliver. But the apparent motion of the Moon also depends on parallax; you are significantly closer to the Moon when it is high in the sky than when it is rising or setting. Further, the Moon is moving in declination as well as right ascension. The result is that no single lunar rate is correct all of the time.

It would be straightforward to have the computer automatically select lunar rate when you are observing the Moon, solar rate for the Sun and planets, and sidereal rate for everything else. I am not aware of a computerized telescope that actually does this.

Having said all this, let me point out that the tracking rates of some of the less expensive computerized telescopes are not particularly accurate; the object may wander all over the field of a high-power eyepiece. Telescopes with servo motors and worm drives (such as the Meade LX200) are more precise.

### 4.6.2 Periodic-error correction (PEC)

When it is tracking the stars, an equatorially mounted telescope moves only in right ascension, so it only needs to run one motor. This makes the tracking inherently smoother.

Still, even the best drive motor has some **periodic error** caused by small irregularities in the gears. During one revolution of the main worm gear – which typically takes 8 minutes – the image shifts east or west, then back to its original position.

The best computerized telescopes offer **periodic-error correction** (**PEC**), which means that small irregularities in the gears can be memorized and corrected by the computer, so that tracking is extremely smooth – but only in equatorial mode. The procedure is to use a crosshairs eyepiece at high magnification and keep a star perfectly centered by making manual corrections for 8 minutes. Thereafter, the computer repeats the same corrections, somewhat smoothed out, at the correct interval. For really accurate training, use a CCD autoguider.

Larger observatory telescopes have periodic-error correction on both axes for maximum smoothness even in altazimuth mode.

### 4.7 Setting circles

Many computerized telescopes have **setting circles** (Figures 4.12, 4.13) just like those on noncomputerized telescopes. The declination circle is permanently fixed in position; the right ascension circle can be rotated to align with the celestial sphere.

The declination circle reads 90° when the telescope is pointed directly away from its base, ready for polar alignment. (This is important; use a bubble level to check it and make adjustments as necessary.) Apart from that, the main use of setting circles is as a "sanity check" on the computer's performance, since they read the absolute position of the tube, not just its relative motion.

To use setting circles, first polar-align as accurately as possible; they work only in equatorial mode. Then go to any convenient star. The declination circle

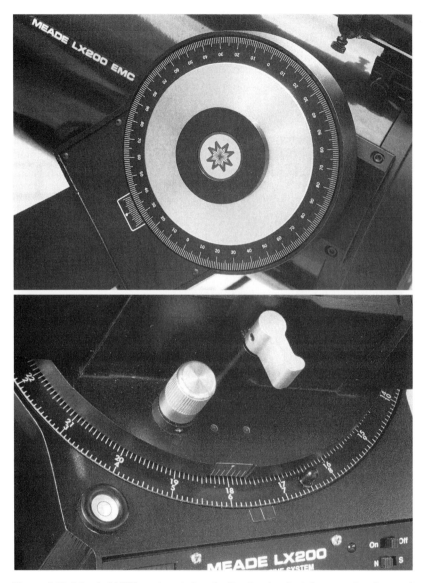

Figure 4.12. Meade LX200 setting circles: declination (top), right ascension (bottom).

should show that star's declination; if it does not, either the declination circle or the polar alignment is inaccurate.

The right ascension circle rotates freely. Set it so that the pointer indicates the star's right ascension. Then ascertain whether the right ascension circle is driven by the telescope's tracking motor. If so (as on the Meade LX200), it will remain accurate throughout the evening; if not (as on the ETX-90), it is accurate only when you have just set it.

The right ascension circle may have two sets of numbers on it, one reading forward (1...2...3...) and one backward (3...2...1...). Assuming you are

Figure 4.13. Meade ETX-90 setting circles: declination (top), right ascension (bottom). LX90 is similar.

north of the Earth's equator, if the circle is attached to the base, as on the LX200, use the scale that reads forward. If the circle moves with the telescope, as on the ETX-90, use the scale that reads backward. When in doubt, set the circle for one star, then go to another star and see if it is correct.

## 4.8  Southern declination limits

At low latitudes, fork mounts in equatorial mode cannot point at the southernmost part of the sky (Figure 4.14). The reason is that a fork-mounted telescope cannot point directly away from the pole; if it did, it would be looking through its own base. Accordingly, there is a limit to how far south the telescope can be aimed, and if you live in the tropics or subtropics, part of the sky may be blocked.

This limit can be expressed as a declination, independently of the observer's latitude. For example, most Schmidt–Cassegrain telescopes experience some blockage when aimed south of declination −50° or −55°, which is not a problem in the continental United States. (At New York, declination −50° is on the horizon.)

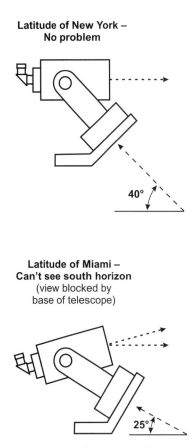

**Latitude of New York –
No problem**

40°

**Latitude of Miami –
Can't see south horizon**
(view blocked by
base of telescope)

25°

Figure 4.14. A fork-mounted equatorial telescope cannot see the far southern sky at low latitudes because the base blocks its view.

Maksutov–Cassegrains, with their longer tubes, have more of a problem. For example, the Meade ETX-90 "bottoms out" at just −35°, which means that even in New York, part of the southern sky is not viewable in equatorial mode.

German-style equatorials do not have this problem, nor do fork mounts in altazimuth mode.

## 4.9 German equatorial mounts

So far we have been considering only fork mounts. German equatorial mounts (GEMs, Figure 4.15) – the kind with the counterweight – can also be computerized. This type of mount should perhaps be called "Estonian" because it was first used on the 9.5-inch (24-cm) refractor in Tartu (Dorpat), Estonia, completed in 1826 – but its inventor was a German, Joseph Fraunhofer.

High-quality computerized German mounts are made by Vixen, Losmandy, Software Bisque (Paramount, shown in Figure 4.16), and Astro-Physics. Unlike forks, German mounts can be sold separately from the telescope because the

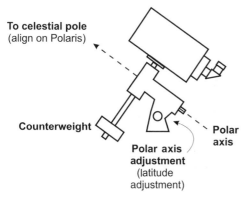

**To celestial pole**
(align on Polaris)

**Counterweight**

**Polar axis**

**Polar axis adjustment**
(latitude adjustment)

Figure 4.15. "German" equatorial mounts overcome the limitations of a fork.

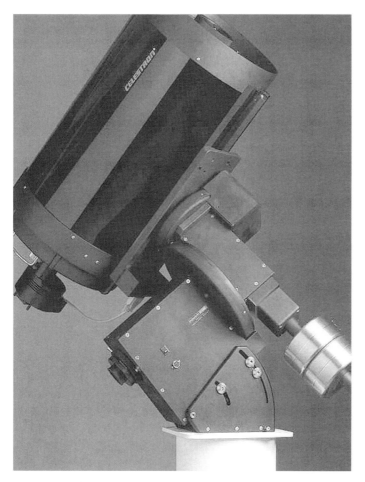

Figure 4.16. The Paramount, an observatory-quality computerized German equatorial mount made by Software Bisque.

57

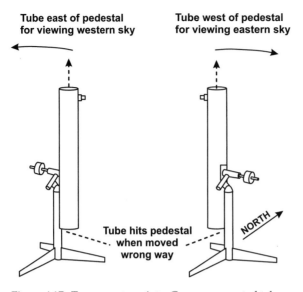

Figure 4.17. Two ways to point a German-mounted telescope straight up. To roll over easily from one position to the other, point the telescope at Polaris.

same mount can be used with telescopes of different sizes. Some amateurs have one excellent mount and several telescopes that can be attached to it.

Setting up a GEM can be easier than polar-aligning a fork mount because the mount is easier to adjust and the finder is never hidden between or under the fork arms. The polar axis is an axle, and you can easily see which way it points; some GEMs even have built-in alignment scopes for sighting Polaris. Ordinarily, though, polar alignment is done by pointing the telescope parallel to its polar axis, then adjusting the mount to point the telescope at the pole.

The GEM has one quirk that can puzzle anyone who hasn't thought it out beforehand. As Figure 4.17 shows, there are two ways to aim the telescope high in the sky, depending on whether you want to have room to swing east or west. To "roll over" easily from one position to the other, aim the telescope directly toward or away from the pole.

# Chapter 5
# Telescope optics

## 5.1 How a telescope works

Figure 5.1 shows the essential parts of a telescope: an **objective** and an **eyepiece** (**ocular**). The objective forms an image, and you use the eyepiece as a magnifier to look at it. The objective and eyepiece each normally have more than one lens element.

The **aperture** of the telescope is the diameter of the objective, which determines the amount of light it collects. This is the most important optical parameter. An 8-inch or 20-cm telescope, for instance, is one with an aperture of 8 inches (20 cm).

The **focal length** of the telescope determines the image size. If the objective is a simple lens, then the focal length is the distance from lens to image. With more complicated telescopes, the focal length may be appreciably more than the physical length of the telescope; it is defined in terms of the simple lens that would give an image of the same size.

The **magnification** (**power**) of the telescope depends on the eyepiece:

$$\text{Magnification} = \frac{\text{Focal length of telescope}}{\text{Focal length of eyepiece}}$$

For example, a telescope with 2000 mm focal length and a 25-mm eyepiece magnifies 80× because 2000 ÷ 25 = 80.

That is why astronomical telescopes are not rated as "10×" or "100×" the way binoculars and microscopes are. By changing eyepieces, you can get any power you want, but not all powers work equally well.

## 5.2 Upside down and backward images

The image in a telescope is normally upside down or reversed left to right. Figure 5.2 shows why. The objective lens flips the image over, but the eyepiece does not flip it again, so if there are no other mirrors or prisms, what you see is

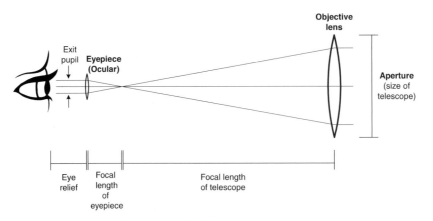

Figure 5.1. Important dimensions of a telescope.

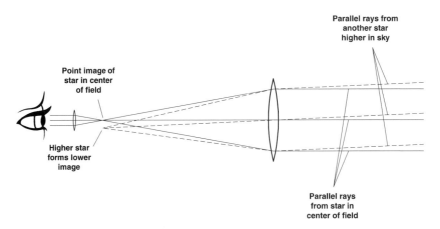

Figure 5.2. Why the image in a telescope is upside down.

inverted. (The eyepiece adds power to the lens of your eye but does not form a separate image of its own.) Extra lenses, prisms, or mirrors could put the image right side up, but they would add cost and bulk, as well as the risk of losing light or reducing image quality. Since "up" and "down" don't mean much in the sky anyhow, astronomers have worked with inverted images for centuries.

But as Figure 5.3 shows, that's not the whole story. Most modern telescopes *do* include another mirror or prism, a **diagonal**, so that observers don't have to crane their necks. The diagonal flips the image, but only in one dimension. Accordingly, what you actually see is *right side up* (because the mirror has flipped it vertically) but *reversed left to right* (because in the horizontal dimension it's still inverted).

Many of the maps in this book are reversed to match such a view. Except for the "R" maps of the American Association of Variable Star Observers, other astronomical publications do not do this. Most star maps are either right side up (north up, east to the left) or inverted (south up, east to the right) to match

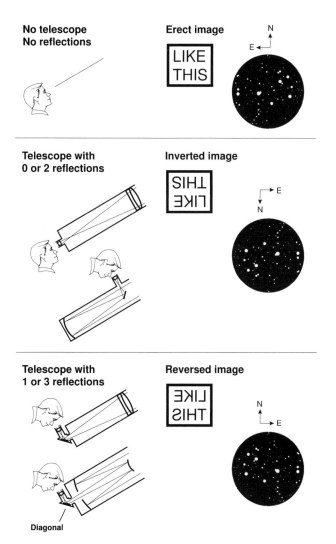

Figure 5.3. Image in telescope may be upside down or reversed left to right.

older types of telescopes. To know for certain which way a map is oriented, you need to know the directions of both north and east.

The diagonal in the Questar and Meade ETX is built-in. Newtonian and Schmidt–Newtonian reflectors don't include a diagonal; they give an inverted image. Image-erecting roof-prism diagonals are described on p. 92.

## 5.3 Light grasp and image brightness

Not only does a telescope make things look bigger, it also gathers more light than the unaided eye.

The pupil of the eye is no more than 7 mm in diameter. An 8-inch (20-cm) telescope is 200 mm in diameter and has $(200/7)^2 = 816$ times as much surface area, so it gathers 816 times as much light. The stars look 816 times as bright as with the naked eye. That's impressive.

The brightness of stars is measured in terms of **magnitude**, a scale derived from an ancient classification of stars as "first class," "second class," and so on. Thus the brightest stars are first magnitude; the faintest that you can see with the naked eye are sixth magnitude; and so on. More precisely, Sirius, the brightest star, is magnitude $-1.4$; Vega is magnitude 0.0; the stars of the Big Dipper are around magnitude 2; and in the dark but humid rural Georgia sky I can see down to about magnitude 6.4.

The magnitude scale is logarithmic, so that if you divide the brightness by $x$, you add $2.5 \log_{10} x$ to the magnitude. We just said that an 8-inch telescope makes the stars look 816 times brighter. That means it makes them $2.5 \log_{10} 816 = 7.3$ magnitudes brighter, and if the magnitude limit of your unaided eye is 6.0, then the telescope extends it to 13.3.

In practice, telescopes do a bit better than this calculation suggests; with an 8-inch (20-cm) telescope under good conditions, you can see down to about magnitude 14.0. The reason is that the telescope can be focused perfectly, and all of the light goes through the highly transparent center of the lens of your eye. Accordingly, here are formulae for computing the magnitude limit from the aperture:

$$\text{Limiting magnitude} = 7.5 + 5 \log_{10} \text{aperture (cm)}$$

$$\text{Limiting magnitude} = 9.5 + 5 \log_{10} \text{aperture (inches)}$$

The central obstruction of a Schmidt–Cassegrain costs you less than 0.2 magnitude because it is only 16% of the total surface area. Table 5.1 shows what to expect, bearing in mind that results depend on the site conditions and the observer.

But telescopes brighten only stars, not extended objects. (An extended object is anything that appears bigger than a point.) The reason is that although the telescope gathers more light than the eye, it also spreads it out over a wider area on the retina. The higher the power, the dimmer the image becomes.

Table 5.1 *Expected magnitude limit vs. telescope size*

| Aperture | | Magnitude limit |
|---|---|---|
| inches | cm | |
| 2.4 | 6 | 11.3 |
| 3.5 | 9 | 12.3 |
| 5 | 12.5 | 13.0 |
| 8 | 20 | 14.0 |
| 12 | 30 | 14.9 |

You can demonstrate this yourself by looking at a distant wall or tree with your telescope in the daytime and trying a variety of eyepieces. Compare the view through the telescope with the naked-eye view. The telescopic image will never be brighter; at high powers it will be considerably dimmer. Likewise, a bright planet such as Saturn can be surprisingly dim when viewed through a telescope at high power.

That doesn't mean the telescope does not help you see faint objects. It's still gathering more light, in total, than your eye could gather by itself. A dim image of substantial size contains more light, and is easier to see, than a tiny speck that is equally dim. That's why tenth-magnitude galaxies aren't visible to the naked eye. Without telescopic aid, the eye doesn't pick up enough light from them to stimulate even one cell on the retina. In the telescope, they are plainly visible, although not bright.

## 5.4 Resolving power

A fundamental law of optics is that *you can't magnify detail that isn't there.* No matter how perfect its optics, no telescope can show an infinite amount of fine detail. Resolution is limited by two things: diffraction and the turbulence of the air.

**Diffraction** is the spreading of light waves as they pass through an opening. You may have seen it demonstrated with tiny pinholes or narrow slits. In a high-powered telescope, even the telescope aperture – several inches in diameter – produces a visible amount of diffraction.

Diffraction makes stars look like disks surrounded by rings rather than perfect points. Figure 5.4 shows a pair of stars, close together, as seen in 4-inch (10-cm) and 8-inch (20-cm) telescopes. The larger telescope forms smaller star images. Diffraction also blurs fine detail on planets; larger telescopes show more detail.

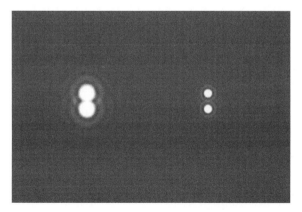

Figure 5.4. A close double star viewed with a 4-inch (10-cm) telescope (left) and an 8-inch (20-cm) telescope (right). Simulated image generated with *Aberrator*.

Table 5.2 *Resolution limits*

| Aperture | Dawes limit | Rayleigh limit (550 nm) |
|---|---|---|
| 60 mm (2.4 inches) | 1.9″ | 2.3″ |
| 90 mm (3.5 inches) | 1.3″ | 1.6″ |
| 12.5 cm (5 inches) | 0.9″ | 1.1″ |
| 15 cm (6 inches) | 0.76″ | 0.93″ |
| 20 cm (8 inches) | 0.58″ | 0.70″ |
| 25 cm (10 inches) | 0.46″ | 0.56″ |

Dawes limit (for double stars that are barely split) and Rayleigh limit (for a somewhat cleaner split) are functions of telescope aperture.

The effect of diffraction on resolving power is expressed as the **Dawes limit** and **Rayleigh limit**:

$$\textbf{Dawes limit} = \frac{4.56''}{\text{aperture (inches)}} = \frac{11.6''}{\text{aperture (cm)}}$$

$$\textbf{Rayleigh limit} \text{ (for 550 nm)} = \frac{5.5''}{\text{aperture (inches)}} = \frac{14''}{\text{aperture (cm)}}$$

The Rayleigh limit is theoretical; the Dawes limit was determined by actual observation of double stars with components of equal brightness. Table 5.2 summarizes these limits for common telescope sizes. For example, a double star with components separated 1 arc-second (1″) is close to the limit of a 5-inch (12.5-cm) telescope but is easily split by an 8-inch (20-cm) in steady air.

If diffraction were the only limiting factor, the largest telescope would always show the best image. But we always observe through the Earth's turbulent atmosphere, which makes the image seethe and boil in a constantly moving blur.

The air is much steadier at some times than others, but even under the best conditions, telescopes larger than about 20 to 30 inches (50 to 75 cm) are unlikely to show additional detail. That's why the Palomar 5-meter (200-inch) telescope is seldom used for planetary work. Under average conditions, the practical aperture limit is more like 8 to 16 inches (20 to 40 cm), and when the air is really rough, even smaller telescopes work best.

## 5.5 Types of telescopes

### 5.5.1 Refractors, reflectors, and catadioptrics

Figure 5.5 shows the main types of telescopes.

The **refractor** uses a two-element lens called an **achromat** to form an aberration-free image. This is a popular design for smaller telescopes and is very rugged.

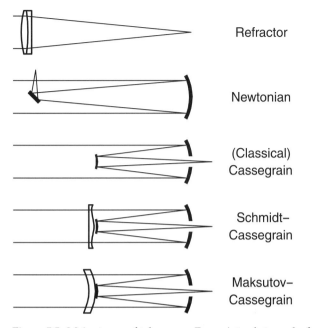

Figure 5.5. Major types of telescopes. From *Astrophotography for the Amateur*, by Michael A. Covington (Cambridge, 1999).

Larger refractors show appreciable color fringing around bright stars unless they use an **apochromatic (apo)** lens that corrects chromatic aberration more effectively. Apo refractors are a relatively new development. They are popular with astrophotographers because they can cover a very wide field – even a full 6 × 6-cm frame of medium-format film – with an image that is sharp all the way to the edges. Apo refractors are neither cheap nor lightweight, and 6 inches (15 cm) is about the practical limit for a portable instrument.

The **Newtonian** (invented by Sir Isaac Newton) is one of several kinds of **reflector** (mirror-based telescope). Newtonians are popular with amateur telescope makers because only one precise optical surface has to be made, the main mirror, which is paraboloidal. Image quality is often excellent and cost is relatively low. Like the refractor, the Newtonian is bulky because the tube length equals the focal length.

Low-cost Newtonians on a simple altazimuth mounts are called **Dobsonians** ("**Dobs**"); they were popularized by John Dobson of the San Francisco Sidewalk Astronomers.

The **classical Cassegrain**, invented by Guillaume Cassegrain in 1672, has a tube much shorter than its focal length. Not only does the folded design save space, but the convex secondary mirror magnifies the image, multiplying the effective focal length. Classical Cassegrains are popular with observatories because the eyepiece is at the lower end of the telescope; you don't have to climb around on ladders the way you do with large Newtonians.

The **Schmidt–Cassegrain** telescope (**SCT**) is even more compact. It is a **cata-dioptric** ("**cat**") telescope, which means it uses both lenses and mirrors.

In its present form, the Schmidt–Cassegrain was developed by Tom Johnson of Celestron in the 1960s. Both mirrors are spherical,[1] hence inexpensive to manufacture.

In front of the secondary mirror is a **corrector plate** that has a slight but complex curvature. Invented by Bernhard Schmidt in 1930, the Schmidt corrector plate corrects the aberrations that would otherwise result from using spherical mirrors.

A **Schmidt–Newtonian** has the mirrors of a Newtonian (except that the primary is spherical) and the corrector plate of a Schmidt. Telescopes of this type are new on the scene and usually have short $f$-ratios; the corrector plate reduces off-axis aberrations.

Another catadioptric design is the **Maksutov–Cassegrain** ("**Mak**" for short), popularized by Questar in the 1950s and by Meade (as the ETX) in the 1990s. Adapted from a 1944 design by D. D. Maksutov, it uses all spherical surfaces and can be made to a high degree of precision.[2] The secondary mirror is an aluminized spot on the corrector plate and does not get out of collimation. Maksutov–Cassegrains are heavier and more expensive than comparable Schmidt–Cassegrains, but they have a reputation for optical superiority.

Maksutov–Cassegrain designs are also used in "mirror lenses" for cameras, but the details of the design are somewhat different. A telescope is designed to form as sharp an image as possible at the center of the field. A camera lens sacrifices some sharpness at the center to get better sharpness at the edges.

## 5.5.2 Catadioptric quirks

Schmidt–Cassegrain and Maksutov–Cassegrain telescopes usually focus by moving the main mirror forward and backward. This movement is amplified by the secondary mirror into a much larger movement of the focal plane. This has advantages and disadvantages.

The obvious advantage is that you can put the focal plane wherever you want it – right at the back of the telescope if you want to place the eyepiece there, or farther back for use with a diagonal, or even a couple of feet past the end of the tube in order to feed a bulky spectrograph or other instrument. Catadioptric telescopes are popular with astrophotographers because they can easily reach focus with any camera body.

---

[1] The original design, published by R. Willey in *Sky and Telescope* (April 1962, pp. 191–193), called for an ellipsoidal secondary, but all indications are that Celestron and Meade use spherical or nearly spherical secondaries.

[2] Meade's 7-inch Mak has an aspheric primary whose exact shape has not been revealed. It is probably very close to a sphere.

The disadvantage is that tiny, unwanted movements of the mirror also have visible effects. It is normal for the image to shift sideways slightly in a random direction when focusing. This **image shift** can be reduced by running the focuser through its whole range several times to redistribute the lubricants. It can be eliminated by using an add-on rack-and-pinion focuser ("zero-image-shift focuser") behind the telescope.

**Mirror shift** – that is, gradual displacement of the mirror as the telescope tilts during tracking – can also affect long-exposure photographs (mostly those longer than 15 minutes) when the tracking is controlled by a separate guide-scope. The remedy is to guide on an off-axis portion of the main telescope's image (p. 118). Meade LX200 GPS telescopes have a lock-down device to hold the mirror in place, and similar devices have been improvised for some other telescopes.

Another catadioptric quirk is that *the focal length changes with focusing.* (After all, that's how a zoom lens works – changing its focal length by changing the separation of elements – and the telescope does the same.) The focal length of a typical 8-inch (20-cm) $f/10$ SCT actually ranges from about 1500 to 2500 mm as the focuser is moved through its range. It has its rated value of 2000 mm with the eyepiece and diagonal in their normal position.

Focusing also affects the correction of spherical aberration. These telescopes are designed to form the best image with the eyepiece and diagonal in their normal position. If you do a star test with the eyepiece right at the end of the tube, you will see more spherical aberration, though the actual degradation of image quality is probably negligible.

### 5.5.3 "Fast" and "slow" $f$-ratios

A "fast" telescope is one with a low $f$-ratio, so that it forms a brighter image and photographic exposures can be completed more quickly. The $f$-ratio is defined as

$$f\text{-ratio} = \frac{\text{focal length}}{\text{aperture}}$$

and has the same significance as the $f$-numbers on a camera lens.

Important as it is for photography, the $f$-ratio is something of a moot point for the visual observer because the image is always viewed through an eyepiece to magnify it further. The ultimate brightness depends on the magnification, not the $f$-ratio.

So which is better, a high or a low $f$-ratio? There are trade-offs. Fast (low-$f$-ratio) telescopes have short focal lengths and are more compact. They give lower-power, wider-field views with common eyepieces.

But optical quality is more of a challenge at low $f$-ratios for several reasons. First, precise manufacturing is more difficult because the lenses or mirrors are more steeply curved.

Second, even with perfect optics, off-axis aberrations are worse. The coma-free image of an $f/10$ Newtonian or classical Cassegrain is over 50 mm in diameter, bigger than the field of any eyepiece. At $f/4$, the image is perfectly sharp only in an 8-mm circle, and even the best eyepiece will not be sharp over a wide field. The 5-meter telescope on Palomar Mountain is $f/3.3$ and requires an extra lens (a Ross corrector) to take photographs on $4 \times 5$-inch plates; without the corrector, it would not even cover 35-mm film.

Third, even with an aberration-free image, a low-$f$-ratio telescope gives its eyepiece a more difficult task. The rays of light are converging more sharply, and a more elaborate eyepiece is needed to focus them correctly, especially at the edges of the field. An $f/15$ telescope works well with any eyepiece, even the lowly Huygenian; an $f/8$ telescope works well with inexpensive Kellners and Plössls; but an $f/4$ telescope may not deliver really sharp images unless you use eyepieces that cost as much as the telescope itself.

In practice, telescopes in the $f/8$ to $f/12$ range are the most versatile, but, by choosing appropriate eyepieces, you can do almost any kind of observing with any telescope. When in doubt, choose a telescope with a higher $f$-ratio because overall optical performance is generally better.

### 5.5.4 Does the central obstruction ruin the image?

One reason apo refractors perform so well is that they have no central obstruction. The central obstruction in a reflector or catadioptric degrades the image – but not as much as some people think.

Why doesn't the obstruction simply produce a hole in the image? Because light from all parts of the image can still reach the focal plane, and no part of the image is affected more than any other.

What the obstruction *does* do is increase diffraction. The light is diffracted by two circular boundaries, not just one. As a result, more light is shifted into the diffraction rings. This causes some blurring of low-contrast planetary detail, but it does not harm the view of double stars or other high-contrast targets (Figures 5.6, 5.7).

The obstruction is normally expressed as a fraction of the aperture. For example, an 8-inch (20-cm) SCT with a 3-inch (7.5-cm) obstruction is slightly less than 40% obstructed.

Unfortunately, misconceptions abound. "A 40% obstruction costs you 40% of the contrast" is simply not true. First of all, it costs you only 16% of the light, because $0.40^2 = 0.16$. Second, removing light does not reduce contrast; the loss of contrast is caused by diffraction, and it only affects details whose contrast was already so low that a small amount of additional light in the diffraction rings can make a difference.

One popular rule of thumb is that if you subtract the diameter of the obstruction from the aperture of the telescope, you get the equivalent unobstructed aperture. Thus an 8-inch telescope with a 3-inch obstruction is as good as a

|       |       |       |
|-------|-------|-------|
| 8-inch (20-cm) telescope<br>No obstruction | 8-inch (20-cm) telescope<br>40% central obstruction | 5-inch (12.5-cm) telescope<br>No obstruction |

Figure 5.6. A close double star viewed in obstructed and unobstructed telescopes. The obstruction shifts more light into the diffraction rings.

 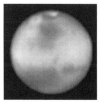 

|       |       |       |
|-------|-------|-------|
| 8-inch (20-cm) telescope<br>No obstruction | 8-inch (20-cm) telescope<br>40% central obstruction | 5-inch (12.5-cm) telescope<br>No obstruction |

Figure 5.7. Simulated views of Mars in perfectly steady air, generated with *Aberrator*. On low-contrast planetary detail, a 40%-obstructed 8-inch (20-cm) telescope (such as the typical Schmidt–Cassegrain) is at least as good as a perfect 5-inch (12.5-cm).

perfect 5-inch, even on the most subtle planetary detail. On double stars and on faint objects requiring larger light grasp, the obstructed 8-inch is of course much better.

If central obstructions degrade the image, why do we put up with them? Mainly because we choose telescopes for cost and portability, not aperture. The choice is not between obstructed and unobstructed telescopes of the same diameter. An 8-inch Schmidt–Cassegrain of average quality outperforms the finest 4-inch refractor – dramatically on many objects, only slightly on some – and is cheaper and more portable. An 8-inch refractor would hardly be portable at all.

### 5.5.5 Which design is best?

Anyone who says that only one type of telescope is respectable, and all the others are junk, should be ignored. All of these telescope designs remain popular because they all work well.

Having said that, I should add that the most consistent reports of high optical quality come from apo refractors, premium-brand Newtonians, and Maksutov–Cassegrains. That is because of a strong tradition of making these kinds of telescopes well, not just the inherent properties of the designs.

Newtonians vary in quality. Some of the least expensive use a spherical mirror instead of a paraboloid. In a 4-inch (10-cm) $f/8$ telescope, that is almost acceptable because the discrepancy is only a quarter of the wavelength of light ("$\frac{1}{4}$ wave" in telescope makers' parlance). Forty years ago, it was widely but incorrectly believed that $\frac{1}{4}$ wave was good enough. The fact is that $\frac{1}{8}$ wave is a more reasonable standard, and higher precision, up to $\frac{1}{20}$ wave or so, pays off under good conditions. Thus, really good Newtonians are much better than the telescopes commonly made by amateurs two generations ago.

The optical quality of Schmidt–Cassegrains improved dramatically in the mid-1990s when both Celestron and Meade retooled their manufacturing processes. The very first Celestrons were reportedly excellent, but during the 1980s, unit-to-unit variation became excessive, and both Celestron and Meade produced some "lemons". Those problems have been solved, and my star tests of recent units, both Celestron and Meade, have consistently shown less than $\frac{1}{6}$-wave total error, equivalent to $\frac{1}{12}$-wave accuracy at the mirror of a Newtonian.

## 5.6 Collimation

Nobody would buy a guitar and not learn to tune it, but, sadly, many Schmidt–Cassegrain owners never learn to **collimate** their telescopes (align the mirrors with the light path), and image quality suffers terribly. To a lesser extent this is also true of Newtonians and has given both kinds of telescopes an undeserved reputation for poor images. Collimating a telescope is much easier than tuning a guitar and every bit as necessary if you want the instrument to work properly.

Most importantly, collimation is nothing to be feared. Even if you can't collimate your telescope perfectly, you can almost certainly improve it, and you're not going to harm it by trying.

### 5.6.1 Collimating a Schmidt–Cassegrain

The only adjustment on a Schmidt–Cassegrain telescope is the alignment of the secondary mirror, located in the middle of the corrector plate. To adjust it, tighten or loosen any of the three adjusting screws (Figure 5.8); the secondary mirror is balanced on a pivot between them.

To do the adjustment, aim the telescope at a star high in the sky, throw it out of focus, and then adjust until the image is a concentric doughnut (Figure 5.9). Rather than try to figure out in advance which screw to turn, just try one of them and see what it does; if it's the wrong one, undo what you did and try another. The image will move when you turn a screw; re-center it before judging the collimation, because the center of the field is where you want the collimation to be best.

In a pinch, if no star is available, you can use a spotlight reflected off a ball bearing on the other side of a large room. Don't try to perform a complete star

Figure 5.8. Collimating a Schmidt–Cassegrain telescope with an Allen wrench.

Figure 5.9. Collimating a Schmidt–Cassegrain. View a star out of focus and adjust until the secondary shadow is centered. Then switch to a higher power eyepiece and try again. Finally, go to Figure 5.13(d) and eliminate any remaining error.

test this way, though, since the telescope is not focused at the distance for which it was designed.

After making a noticeable improvement, bring the image closer to focus or switch to a higher-power eyepiece and try again. When you think the collimation is perfect, compare the in-focus image at high power to Figure 5.13(d). Chances are, you can see and correct a small remaining error. The ultimate accuracy you can achieve will be limited by the steadiness of the air.

Work with very small increments. A tenth of a turn usually makes a noticeable difference. When making larger changes, the rule is to tighten when possible, but if a screw is already tight, loosen the others instead. Never tighten a screw with the full force of the wrench; the screws should be finger tight or slightly

Figure 5.10. Using a ruler to check that the secondary holder is centered. Note thumbscrews installed in place of original collimation screws.

tighter. Never loosen all three screws at the same time, or the secondary mirror may come off its support.

The adjustment normally requires an Allen wrench, but you can install thumbscrews (visible in Figure 5.10) so that you can collimate by hand. One supplier of suitable thumbscrews ("Bob's Knobs") is Morrow Technical Services, 6976 Kempton Rd., Centerville, IN 47330, U.S.A., http://www.bobsknobs.com. When installing them, never remove more than one screw at a time, or the secondary may fall off its support.

All this is not as complicated as it sounds. I can collimate a Schmidt–Cassegrain in two or three minutes, and even a beginner should finish in ten minutes. Much time is saved by not trying to predict which screw to turn; just trying them is a lot faster.

### Is the secondary centered?

If collimation seems unduly difficult or if some astigmatism (Figure 5.14(b)) remains after your best efforts, take a ruler and see if the secondary holder is actually in the center of the corrector plate (Figure 5.10). On the NexStar 5 and possibly other telescopes, the plastic mount of the secondary mirror is held in place by tension in an oversized hole in the corrector plate. The corrector plate itself may also be loose in its mount. Either element can shift sideways, producing a telescope that is seemingly impossible to collimate. This is a relatively rare problem, but I've seen it in two telescopes.

The cure is simple – just push the secondary holder back into position. You can move it laterally by applying gentle pressure with your thumb and fingers. Centering it to within 0.5 mm should be good enough. It doesn't much matter whether you are moving the secondary holder or the whole corrector plate; whatever is slightly loose will move, and slight decentering of the corrector plate is not serious as long as the secondary mirror is centered.

### 5.6.2 Collimating a Newtonian

Collimating a Newtonian is more complicated, but many times, you can skip the complex part and treat it like a Schmidt–Cassegrain – just adjust the three screws on the back of the main mirror until out-of-focus star images look like concentric doughnuts. Even if you don't get perfect collimation, you can often get a substantial improvement.

Complete collimation of a Newtonian is done in the daytime rather than on a star at night. Aim the telescope at a bright surface, such as the daytime sky, and look into the eyepiece tube. To ensure that your eye is centered, punch a hole in the bottom of a 35-mm film can and use it in place of an eyepiece.

Figure 5.11 shows what you should see. You're looking straight at the secondary mirror, which is *not* necessarily concentric with the eyepiece tube (Figure 5.12 shows why).

In the secondary mirror you see a reflection of the primary mirror. Your first job is to adjust the secondary, if it's adjustable, so that the reflection of the primary is centered in the eyepiece tube. Then adjust the primary mirror until the central obstruction is centered in it. There – your telescope is collimated. You can and should still check it on a star and make fine adjustments to the primary mirror if any small errors remain.

All this assumes your telescope was fairly close to correct collimation in the first place. If you're building a Newtonian or repairing one that has been assembled incorrectly, you may need more guidance. The definitive handbook on collimating Newtonians is *New Perspectives on Newtonian Collimation*, by

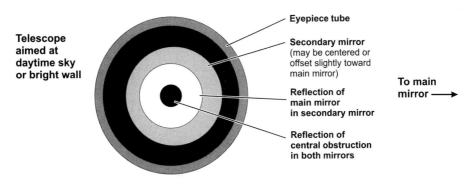

Figure 5.11. Looking down the eyepiece tube of a Newtonian, everything should be concentric except the secondary mirror. Often, the main mirror mount is all you need to adjust.

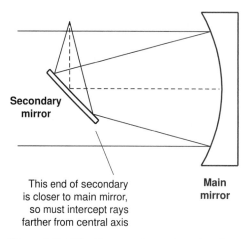

This end of secondary
is closer to main mirror,
so must intercept rays
farther from central axis

Secondary
mirror

Main
mirror

Figure 5.12. Why the secondary mirror is not concentric with the eyepiece tube. In high-$f$-ratio telescopes ($f/8$ and up) it may be concentric after all.

Vic Menard and Tippy D'Auria (privately printed, but available from Sky Publishing Co., Cambridge, Massachusetts 01238, U.S.A., http://www.skypub.com). This book also discusses Schmidt–Cassegrains, conventional Cassegrains, and mirror star diagonals.

If you collimate a Newtonian frequently, you may want to use one of the **laser collimators** that have recently appeared on the market. The idea is that a laser beam, aimed at the primary mirror from the eyepiece position, will bounce back along its path if the collimation is correct. This can be a real time-saver.

### 5.6.3 How often?

How often should a telescope be collimated? As often as it needs it. Collimation should be *checked* by looking critically at star images whenever you use the telescope. (Remember that collimation can only be judged at the center of the field.) With gentle treatment, a Schmidt–Cassegrain that is used at the observer's home can stay collimated for months. If a telescope won't hold collimation – if the collimation shifts during routine transportation or even during an observing session – the collimating screws are probably too loose.

A telescope that has been transported to a remote site may need touching up when it gets there. Shipping the telescope to a dealer for collimation is a dubious idea, since it is likely to shift slightly during the return trip.

### 5.7 Star testing

A star, seen from many light-years away, is an almost perfect source of parallel light rays. By viewing a star at high power under steady air, you can perform very sensitive tests of the quality of your telescope's optics.

(a) Perfect optics: inside focus, in focus, and outside focus.

(b) Same, in unsteady air (the usual situation).

(c) Perfect optics with a thermal air current inside the tube because the telescope has not finished cooling down.

(d) Perfect optics, out of collimation.

Figure 5.13. Star tests of a good telescope under a variety of conditions. (Images generated with *Aberrator.*)

Figure 5.13 shows what a star test looks like. The first picture shows a star as seen with a perfect telescope, inside focus, perfectly focused, and outside focus. The second picture shows a star in turbulent air – the usual situation – and the third shows the effect of a **tube current**, an air current inside the telescope tube. (Besides bashing in one side of the round image, tube currents can also produce "pie-slice" deformations.) The last picture shows miscollimation. All of these are problems that are easy to remedy; let the telescope cool down to match the surrounding air (or wait for better conditions) and adjust the collimation.

(a) Spherical aberration (relatively common).

(b) Astigmatism (uncommon, usually caused by decentered element).

(c) Mirror pinched by three supporting clamps or screws.

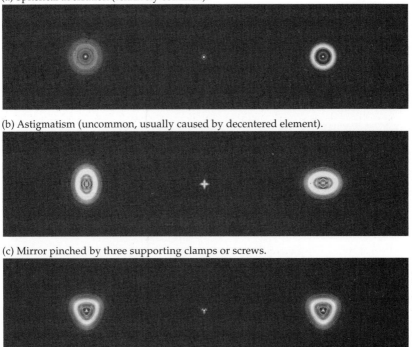

Figure 5.14. Star tests that indicate optical problems. (Images generated with *Aberrator.*)

Real optical defects are shown in Figure 5.14. Spherical aberration makes the out-of-focus image look different depending on the direction in which it is defocused. Small amounts of spherical aberration are common in mass-produced telescopes and have very little effect on image quality. Spherical aberration is often temperature-dependent, influenced by the expansion and contraction of the main mirror.

Astigmatism is what happens when an element has elliptical rather than circular symmetry. It is a relatively rare problem in telescopes but common in human eyes, so the first thing you should try is tilting your head to see if the astigmatism tilts with you. If so, it's in your eye, and you should wear glasses while observing. Otherwise, look for a decentered element (p. 72).

A triangular kind of astigmatism happens when a mirror is pinched by its three mounting screws or clips, and small amounts of this problem are sometimes visible even in very well-made telescopes. A tube current can mimic this problem.

The images in these pictures were generated with *Aberrator*, a free computer program distributed by its author, Cor Berrevoets (http://aberrator.astronomy. net). You can determine the exact condition of your telescope's optics by finding the parameters that enable *Aberrator* to match it. The definitive manual of star testing is *Star Testing Astronomical Telescopes*, by H. R. Suiter (Willmann-Bell, 1994), which includes numerous computer-generated illustrations.

But a word of caution is in order. Star testing may be *too* sensitive. It can reveal paltry defects that have no visible effect on the image quality. Moreover, you are at the mercy of atmospheric conditions and thermal equilibrium. A single star test can (with luck) prove that a telescope is good, but a single test cannot prove that a telescope is bad, unless a defect is very prominent. Normally, tests must be repeated over and over at different temperatures, under different atmospheric conditions, before a verdict can be given.

There are no perfect telescopes, only telescopes whose defects have not yet been measured. Some large observatory telescopes have notorious defects. The relevant question is not, "Is it perfect?" but rather, "Can you do astronomy with it?" The answer is almost always yes.

## 5.8 Buying a telescope secondhand

Until computers came along, a telescope was, for all practical purposes, an instrument that would never wear out or become obsolete. Accordingly, amateur astronomers have long relied on being able to buy telescopes secondhand and resell them later at little or no loss. Today, secondhand telescopes are available through many dealers; privately, by contacting members of astronomy clubs; and through websites such as Astromart (http://www.astromart.com), Astronomy-Mall (http://www.astronomy-mall.com) and eBay (http://www.ebay.com).

Computerization introduces three wrinkles into the situation. First, the complex mechanisms are somewhat failure-prone and subject to wear. Second, electronic failures, when they occur, are often repairable only by the manufacturer. Third, and most seriously, firmware is almost always imperfect, and other aspects of the design are likely to be, at this early date, unproven. Accordingly, an older telescope is likely to be inferior to a newer one of the same model, which will incorporate minor, unannounced engineering changes.

There are ways to get around this. New firmware for the Meade Autostar can be downloaded through the Internet and installed from your PC. Updated electronics for early-model LX200s are available from Meade. Finally, of course, if a telescope performs satisfactorily, there is no reason to reject it just because it's an early model. Plenty of early model Ultima 2000s, LX200s, and ETX-90s will remain in circulation and in use for decades.

Indeed, in one respect a slightly used telescope may be better than a new one. Manufacturing defects are most likely to appear during the first few weeks of use. A telescope that gets through its break-in period, and continues to be treated well, is generally reliable for years to come.

## 5.9  Cleaning optics

### 5.9.1 Lenses

There is no need to clean a refractor objective or a Schmidt–Cassegrain corrector plate until a noticeable accumulation of dust has developed. A few specks do not count. Except for blowing dust off, no cleaning may be needed for years. Eyepieces need cleaning much more often, since they accumulate grease spots that look like defects in the anti-reflection coating.

The best way to remove dust is to use compressed gas, such as "Dust-Off," or a very soft brush. When a telescope lens must be wiped to remove grease, you can use any method that is safe for camera lenses. Two products that I find particularly useful are PEC-Pads (lint-free wipes) and Eclipse (a methyl-alcohol-based lens cleaner), both made by Photographic Solutions, Inc., P.O. Box 135, Onset, MA 02558, U.S.A. (http://www.photosol.com), and widely available in camera stores. Microfiber cloth also works well, as do very well-laundered, nearly worn-out handkerchiefs. High-quality tissue paper is useful if completely free of grit, oil, and perfume, but I have never had much use for old-fashioned lens tissue, which is too thin.

Never apply liquids directly to the lens; they will run under its edges and cause trouble. Instead, apply a drop or two of liquid to the cloth. Suitable lens cleaners include various mixtures of pure water; isopropyl, ethyl, or methyl alcohol (*without* impurities that would leave a residue); and ammonia-based cleaners such as Kodak Lens Cleaner and even Windex.

Do not use "anti-fog" lens cleaners; they leave a residue that prevents the anti-reflection coating from working.

The main danger when wiping a lens is that a particle of grit will cause a scratch. For that reason, always remove dust with compressed gas or a brush before wiping, and avoid repetitive circular motions.

### 5.9.2 Mirrors

Never rub a mirror and never use an ammonia-based cleaner on it, or the aluminum coating may come right off. When a mirror needs cleaning – as happens every few years with Newtonians, since there is no good way to keep dust off the mirrors – use a very dilute solution of dishwashing liquid in distilled water, just a few drops per liter. Remove dust and grime with a stream of liquid, not by rubbing. If you must rub the surface, use cotton balls and be *extremely* gentle.

# Chapter 6
# Eyepieces and optical accessories

## 6.1 What eyepieces do you need?

The ease and comfort of using a telescope depend to a surprising degree on the eyepiece, which contains more optical elements than the rest of the instrument. Most importantly, the eyepiece determines the magnifying power.

The magnifications that work well with any telescope are proportional to its aperture. For example, a 4-inch (10-cm) telescope at 25× and an 8-inch (20-cm) at 50× are both at the low end of their power ranges. They are operating at 2.5 power per centimeter of aperture, or 2.5×/cm for short, equivalent to about 6 power per inch.

What this implies is that the eyepiece focal lengths that work well with any telescope depend on its $f$-ratio. For example, a 20-mm eyepiece gives low power with an $f/5$ telescope but medium power with an $f/10$ telescope.

With that in mind, Table 6.1 makes specific recommendations. You do not have to get the exact focal lengths in the table, of course, but the table will show you where any particular eyepiece stands. If you have an $f/10$ telescope and choose to get a 32-mm eyepiece, which is not in the table, you'll know that it's in the low- to very-low-power range.

You do not need a lot of eyepieces. Three eyepieces are enough for almost anybody; two (low and high power) are almost enough; and one eyepiece (low or medium power) will get you started. Go for quality, not quantity.

## 6.2 Barrel size

Naturally, the eyepiece must fit your telescope. Four barrel sizes (tube sizes) are in common use:

- 23.3 mm (0.917 inch), used on microscopes;
- 24.5 mm (0.965 inch), used on smaller Japanese telescopes;

Figure 6.1. Eyepieces. Back row, left to right: 40-mm König (an Erfle variant) in a 2-inch barrel; 14-mm Radian and 9-mm Lanthanum LV in $1\frac{1}{4}$-inch barrels. Front row: 18-mm orthoscopic with parfocalizing ring held in place by set screw; 25-mm Plössl; Celestron Ultima 2× Barlow lens.

- $1\frac{1}{4}$ inches (32 mm), standard for most telescopes nowadays; and
- 2 inches (50.8 mm), used on larger telescopes and long-focal-length eyepieces.

Four-inch (100-mm) eyepiece tubes are occasionally used at observatories.

Why so many sizes? Partly for historical reasons, and partly because long-focal-length, wide-angle eyepieces require big tubes. The lenses of a 6-mm eyepiece will fit in any size tube, but a 40-mm eyepiece with a decently wide field needs a 2-inch tube. I have a 50-mm eyepiece in a 23.3-mm (microscope-standard) tube, but its field of view is quite narrow; it's almost like looking through a drinking straw.

You can generally use more than one size of eyepiece with the same telescope. On most Schmidt–Cassegrains or Maksutov–Cassegrains, you can use either a 2-inch or a $1\frac{1}{4}$-inch eyepiece holder. Many Newtonians have a 2-inch eyepiece tube with a removable $1\frac{1}{4}$-inch bushing. With Japanese-size (0.965-inch) refractors, you can use a "hybrid diagonal" that takes $1\frac{1}{4}$-inch eyepieces.

## 6.3 Field of view

The **apparent field** of an eyepiece is its field of view as seen by the observer's eye. This contrasts with the **true field**, which is the amount of sky you are actually

Table 6.1 *Recommended eyepiece focal lengths*

| Power | Used for | Exit pupil | Power per cm (per inch) | Telescope $f$-ratio | | | | | |
|---|---|---|---|---|---|---|---|---|---|
| | | | | $f/4.5$ | $f/6$ | $f/8$ | $f/10$ | $f/12$ | $f/15$ |
| Very low Note 4 | Nebulae, galaxies under very dark skies | 4–7 mm | 1.4–2.5×/cm (3.5–6×/inch) | 25 mm Note 1 | 32 mm | 40 mm Note 3 | 40 mm, or 25 mm with focal reducer Note 3 | 50 mm, or 32 mm with focal reducer Note 3 | 60 mm, or 32 mm with focal reducer Note 3 |
| Low Note 4 | Nebulae, galaxies, star clusters, finding objects | 2–4 mm | 2.5–5×/cm (6–12×/inch) | 12 mm Notes 1, 2 | 17 mm | 22 mm | 25 mm | 32 mm | 40 mm Note 3 |
| Medium | General observing, Moon, planets | 1–2 mm | 5–10×/cm (12–25×/inch) | 6 mm Notes 1, 2 | 9 mm Note 2 | 12 mm Note 2 | 15 mm | 17 mm | 20 mm |
| High | Moon, planets, double stars in very steady air | 0.5–1 mm | 10–20×/cm (25–50×/inch) | 3 mm, or 6 mm with 2× Barlow Notes 1, 2 | 4 mm Note 2 | 6 mm Note 2 | 8 mm Note 2 | 9 mm Note 2 | 10 mm Note 2 |

*Notes:*

1 At $f$-ratios below 6, classic eyepiece designs such as Orthoscopic and Plössl may fail to give a sharp image. Modern designs are better; if possible, the eyepiece should be tried in the telescope.

2 Below 15 mm (25 mm for eyeglass wearers), long-eye-relief designs are preferred.

3 A 2-inch tube is preferable for 40-mm eyepieces and mandatory for longer focal lengths.

4 Low powers are not usable in the daytime in centrally obstructed telescopes because the pupil of the observer's eye is not large enough to clear the obstruction in the exit pupil.

Figure 6.2. Same magnification, different fields of view. Left: 50 power, 35° apparent field, 0.7° true field. Right: 50 power, 60° apparent field, 1.2° true field. (For realistic views, hold the page close to your face.)

looking at (Figure 6.2). For example, the telescope might take a 1° circle of sky and magnify it 30×, so that it appears to be 30° across.

In a distortion-free eyepiece,

$$\text{True field} = \frac{\text{Apparent field}}{\text{Magnification}}$$

but in practice, some eyepieces have a slightly different magnification near the edges than at the center, and this relation is not exact.

There are two reasons why a wider field is better, at least up to a point. One is comfort. The human eye naturally takes in a field of about 50° or 60°, and eyepieces with fields in this range are comfortable to use. Narrower fields, such as the 40° that used to be standard, give the impression of looking through a round window.

The other is pointing accuracy. With a wide-field eyepiece, an imperfectly located object will still be in the field even at medium or high power. You must still center the object, since the image is sharper in the center, but at least you can see it without changing eyepieces.

### 6.3.1 Measuring field of view

Here is a simple method to measure the true field of view of any telescope:

(1)    Choose a star whose declination is within a few degrees of 0° and center it.
(2)    Turn off the drive.
(3)    Time how long it takes the star to drift out of the field.

The field of view, in arc-minutes, is the time in seconds divided by 2. (One arc-minute corresponds to four seconds of time, but you are timing the star's passage across only half the field.)

## 6.4  Eye relief

The distance from the eyepiece to your eye is called the **eye relief**. If you wear glasses while observing, the eye relief should be at least 15 and preferably 20 mm; even without glasses, you'll probably find eye reliefs shorter than 10 mm uncomfortable.

An eyepiece has slightly more eye relief in the telescope than by itself. That is, if you pick up an eyepiece and look through it in the showroom, you'll have to put your eye slightly closer to see its full field than when using it on a telescope. Adding a Barlow lens increases the eye relief even further, but the change is not large.

Some eyepieces have so much eye relief that you're likely to put your eye too close. When that happens, the outer part of the image disappears. I've used 55-mm Plössls that have enough eye relief for an elephant. In one case I made a small cylinder of black foam (from an art supply store) to use as an improvised eyecup.

The sensation that you can't get your eye into the right position can also be caused by the shadow of the central obstruction when you try to use low power in the daytime (p. 88).

## 6.5  Eyepiece designs

After a century of stagnation, eyepiece technology began to advance rapidly around 1980, and today you can buy finer eyepieces than previous generations would have dreamed of. You can also spend more money on them. There have been three main advances: sharper images, more eye relief, and wider fields.

Figure 6.3 shows a range of eyepiece designs. At the bottom is the 300-year-old **Huygens** (*HOY-khens*) or **Huygenian** (**H**) design, a cheap two-element eyepiece still supplied with some low-cost telescopes. It works well only at high $f$-ratios and I do not recommend it. The improvement when switching to a better eyepiece, even a humble Kellner, is often dramatic.

The **Kellner, Achromatic Ramsden**, and **Modified Achromatic** (**K, AR**, and **MA** respectively) are three variations on a 150-year-old three-element design. This was the usual general-purpose eyepiece a generation ago; today, most people go for Plössls and Orthoscopics, which are sharper and give more eye relief.

The **Abbé Orthoscopic** has a following among planet observers and until the 1970s was generally considered the best eyepiece available. The images are sharp and eye relief is relatively abundant.

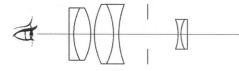

**Radian**
Al Nagler, 1999
Apparent field 60°
Eye relief 20 mm
Diagram is an educated guess;
design has not been published.

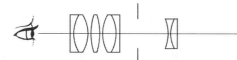

**Lanthanum LV**
Vixen Ltd., 1997
Apparent field 50°
Eye relief 20 mm

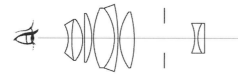

**Nagler Type 2**
(one of several types)
Al Nagler, 1987
Apparent field 82°
Eye relief about 10 mm

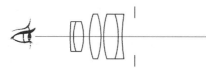

**Erfle (Er)**
(one of several variations)
Heinrich Erfle, 1917
Apparent field 65°
Eye relief 35% of focal length

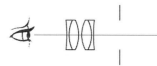

**Plössl (Pl)**
Georg Simon Plössl, 1860
Improved by Nagler, 1979
Apparent field 50°
Eye relief 80% of focal length

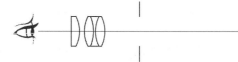

**Orthoscopic (Or)**
Ernst Abbé, 1860
Apparent field 45°
Eye relief 80% of focal length

**Kellner (K)**
**Achromatic Ramsden (AR)**
**Modified Achromatic (MA)**
Carl Kellner, 1849
Apparent field 40°
Eye relief 40% to 50% of focal length

**Huygens (Huygenian) (H)**
Christiaan Huygens, 1703
Apparent field 35°
Eye relief 30% of focal length

Figure 6.3. Some popular eyepiece designs, with the newest at the top.

In 1979, Al Nagler of Tele Vue Optics invented a minor improvement to the century-old **Plössl** (*PLUS-l*) design, and Tele Vue Plössls quickly became the standard to which other eyepieces were compared. Celestron, Meade, and other manufacturers responded by marketing their own Plössls. Today, the Plössl is the workhorse of eyepieces, the usual choice when other concerns are not overriding. No other type gives a sharper image at the center of the field. To get a slightly wider field of view, several manufacturers add a fifth lens element.

Eyepieces of the **Erfle** type are popular with observers who enjoy wide fields.[1] Many observers first learned about the Erfle via surplus eyepieces from World War II gunsights.

## 6.6 New-generation eyepieces

At the top of Figure 6.3 are three examples of new-generation eyepiece design, a Tele Vue Radian, a Vixen Lanthanum LV, and a super-wide-field Nagler Type 2.

The Radian (from Tele Vue) and the Lanthanum LV (from Vixen, also sold by Celestron and by Orion in the U.S.A.) provide 20 mm of eye relief regardless of focal length. For many eyeglass wearers, including myself, eyepieces like these are what make planetary astronomy possible. Before they came out, I seldom viewed the planets because I could never get my eye close enough to a 9-mm Orthoscopic or Plössl.

The Nagler Type 2 is one of a series of eyepieces that have extremely wide fields. They do not accommodate eyeglass wearers, but without glasses, they are quite comfortable and give dramatic views of star clusters and nebulae.

These designs and many others like them are the fruits of three technological advances. Better anti-reflection coatings made it practical to use more lens elements. New types of glass, including lanthanum glass, gave the designer more control over aberrations. Most importantly, computers made it easy to calculate exactly how an eyepiece will work, so designers no longer had to stick to small variations on classic designs.

The light enters all these new designs through a negative (concave) lens group that acts like a built-in Barlow lens (p. 92). This magnifies the image so that the rest of the eyepiece can have a longer focal length and more eye relief. It also narrows the cone of light so that the eyepiece works well even with telescopes of low $f$-ratio. After the negative element, the rest of the eyepiece resembles a Plössl (in the Radian and Lanthanum LV) or something in the Erfle family (Nagler).

---

[1] English speakers usually pronounce *Erfle* as *UR-fle*, with the second syllable as in *rifle*, but the name is actually German and was originally pronounced *EHR-fla*.

## 6.7   Anti-reflection coatings

All well-made lenses since the 1950s have had anti-reflection coatings.[2] These coatings serve two purposes. They keep light from being lost, and they prevent ghost images or flare (scattered light) that would otherwise result from reflections.

The latter function is more important. A small amount of lost light won't be missed; the human eye can't even detect a loss of less than about 10%. But scattered light can hide faint objects, and bright reflections can be annoying or even misleading; in the 1700s, reflections were mistaken for a satellite of Venus.

Since the 1970s, the best camera lenses and eyepieces have been **multi-coated** to reduce reflections to the lowest level possible. Multi-coating is useful but not essential; single coating can be almost as good, especially in a lens with only a few elements. (Uncoated lenses are rare and should be avoided.) Stray light can also be reduced by blackening the edges of the lens elements.

Damage to lens coatings is, in my experience, rare, but grease can make the coating appear to be badly damaged or nonexistent. See p. 78 for advice on cleaning optics.

## 6.8   Choosing eyepieces wisely

The most expensive eyepieces are not necessarily the best; it depends on what you are paying for. It's wise to put image quality first, comfort second, and wideness of field third.

Telescopes faster than $f/6$ probably need new-technology eyepieces at all focal lengths. Otherwise, Plössls and Orthoscopics give as sharp an image as anything else and are much less expensive. All well-established brands are well made. Even a Kellner can be quite satisfactory as a 32-mm low-power eyepiece.

At higher powers you will probably want more eye relief than a Plössl can give. Here the Lanthanum LV and Radian stand out. (Of the two, the Lanthanum LV is considerably less expensive.) If you don't wear glasses, a variety of other designs are also suitable.

Deep-sky observers who enjoy a really wide field can invest in Naglers, Meade and Vixen superwides, and other extremely wide-field eyepieces. Personally, I've never gotten used to them. If the eyepiece shows more than I can take in at once, I feel I'm not getting full value from the eyepiece.

It's convenient for all of your eyepieces to be **parfocal**, that is, to focus at the same or nearly the same position, so that you can change eyepieces without refocusing. My Plössl and Radians are designed to match, and my 40-mm König has had a rubber ring added to make it roughly parfocal with them.

---

[2] Except Schmidt–Cassegrain corrector plates, which were uncoated until about 1980. The manufacturers felt there was no danger of internal reflections.

To get a smooth progression of powers, choose eyepieces whose focal lengths form a *geometric progression.* That means each value is some fixed multiple of the previous one. For example, the sequence 5 mm, 10 mm, and 20 mm, in which each focal length is double the previous one, is more useful than 5, 10, 15 or 10, 20, 30. To fill in the gap between two values, take the square root of their product. Thus the gap between 10 mm and 40 mm is best filled by a 20-mm eyepiece because $\sqrt{40 \times 10} = 20$. Where the eyepieces are of different types, Al Nagler suggests making a geometric progression of field stop diameters (p. 89) rather than focal lengths.

## 6.9 Eyepiece calculations and technical details

### 6.9.1 The exit pupil

The exit pupil is the bundle of light rays emerging from the eyepiece into the observer's eye (Figure 5.1, p. 60). You can see it as a bright disk viewed from a few inches away (Figure 6.4). Its size depends on aperture and magnification:

$$\text{Diameter of exit pupil} = \frac{\text{Telescope aperture}}{\text{Magnification}}$$

Figure 6.4. The exit pupil is the bright spot that you see in the middle of the eyepiece.

For example, an 8-inch (20-cm) telescope at $100\times$ has an exit pupil diameter of:

$$\frac{20 \text{ cm}}{100} = 0.2 \text{ cm} = 2 \text{ mm}$$

Higher powers give smaller exit pupils, and any given exit pupil size corresponds to a fixed "power per inch" or "power per cm" of aperture:

$$\text{Magnification} = \frac{\text{Aperture}}{\text{Exit pupil diameter}} = \frac{1}{\text{Exit pupil diameter}} \times \text{Aperture}$$

For example, a 0.1-cm exit pupil corresponds to a magnification of 10 times the aperture in centimeters ("$10\times$ per cm").

Exit pupil size can also be found from $f$-ratio and eyepiece focal length:

$$\text{Exit pupil diameter} = \frac{\text{Eyepiece focal length}}{\text{Telescope } f\text{-ratio}}$$

Eyepiece focal length = Exit pupil diameter $\times$ Telescope $f$-ratio

For example, a 32-mm eyepiece gives a 3.2-mm exit pupil with any $f/10$ telescope.

## 6.9.2 Limits on low power

If the exit pupil is larger than the pupil of the eye, some light will be wasted. That's why the image does not continue to get brighter as you reduce the power below whatever will give a 7-mm exit pupil with your telescope – that is, about 1.4 times the aperture in centimeters.

In practice, the outermost parts of the lens of the eye are of poor optical quality, so the largest useful exit pupil for really sharp images is generally about 5 mm. This makes the lowest useful power about 2 times the aperture in centimeters.

With an unobstructed telescope, you can use even lower powers as long as you don't mind losing some light. The resulting wide-field views are useful and enjoyable even if not theoretically advantageous.

But when there is a central obstruction, you run into another limit. The obstruction causes a hole in the middle of the exit pupil. That hole must be appreciably smaller than the pupil of the eye, or there will be a black shadow in the middle of the image.

That's why low powers on obstructed telescopes do not work well in the daytime. In daylight, the pupil of the eye may be as small as 1 mm. If there is also a 1-mm hole in the exit pupil, the observer will come away with the uneasy feeling that no position of the eye is entirely correct. At night, the same eyepiece works beautifully.

Using a wide-angle eyepiece in the daytime sometimes reveals another problem, the so-called **kidney bean effect.** A bean-shaped shadow, always off center, seems to dart around the field as you move your eye. This is caused by a nonuniform exit pupil. The exit pupils from different parts of the image do not all coincide, and your eye cannot take them all in at the same time. This problem also disappears at night, when the pupil of the eye is wider.

### 6.9.3 Limits on high power

There is no reason to use more magnification than the minimum sufficient to see all the detail that the telescope can show. If you try, all you'll get is a dim, blurry image, badly distorted by diffraction and atmospheric turbulence. You will see more at a lower power.

The practical upper limit is usually about 10 times the aperture in centimeters, corresponding to an exit pupil of 1 mm. This is not a hard and fast limit. Experienced observers sometimes use higher powers; beginners are more comfortable with lower powers.

Depending on the site, there is an absolute limit of about 300× to 1000× because of the turbulence of the air, regardless of the size of the telescope. This creates a curious situation with large observatory telescopes – the lowest useful power and the highest useful power are often the same. Even moderately large telescopes experience this to some extent. The University of Georgia's 24-inch (60-cm) Cassegrain is surrounded by rather turbulent air, so it is almost always used at 200×, regardless of the object being viewed.

Very small exit pupils reveal **floaters** (moving opaque particles) and other imperfections in the human eye. Floaters can be quite annoying; that is one reason I usually use an exit pupil larger than 1.5 mm. They go away completely with a binocular viewer because each eye fills in gaps in the other eye's image. That is why binocular heads are so often used on microscopes, which have tiny exit pupils.

### 6.9.4 Field stop and tube size

The **field stop** is the ring, within the eyepiece, that defines the edges of the image. It is depicted as a pair of short vertical lines in Figure 6.3 (p. 84). Naturally, its diameter must be smaller than the eyepiece tube.

Assuming no distortion, the relation between field stop size, apparent field, and focal length is

$$\text{Field stop size} = 2\tan\frac{\text{Apparent field}}{2} \times \text{Focal length}$$

This is actually the *effective* field stop size, ignoring the effect of any lenses that intercept the light and change the image size before it reaches the field stop.

From the formula, you can see that a 40-mm eyepiece with 60° apparent field must have a field stop diameter of

$$2\tan\frac{60°}{2} \times 40\text{ mm} = 2\tan 30° \times 40\text{ mm} = 2 \times 0.577 \times 40\text{ mm} = 46.2\text{ mm}$$

which is too big for a $1\frac{1}{4}$-inch tube, so a 2-inch tube is required.

If you know the size of the field stop, you can calculate the true field of an eyepiece regardless of distortion. The relation is

$$\text{True field} = 2\arctan\frac{\text{Field stop diameter}}{2 \times \text{Telescope focal length}}$$

For example, the field stop of a 14-mm Radian eyepiece is 14.4 mm in diameter. With a telescope whose focal length is 2000 mm, it gives a true field of

$$2\arctan\frac{14.4}{2 \times 2000} = 2\arctan 0.0036 = 2 \times 0.206° = 0.412°$$

A simpler formula that is accurate enough for almost all purposes is

$$\text{True field} = \frac{\text{Field stop diameter}}{\text{Telescope focal length}} \times 57.3°$$

The latter formula gives 0.413°. Considering that the specified focal length is probably accurate only to within several percent, that is good enough.

## 6.10 Eyepiece accessories

### 6.10.1 Diagonals

A **star diagonal** (Figure 6.5) is a right-angle adapter, now almost universally used with refractors and catadioptrics so that observers do not have to bend their necks backward. It is called a star diagonal to distinguish it from a **sun diagonal**, which reflects only part of the light and was formerly used by solar observers. Some Maksutov–Cassegrains, such as the Questar and Meade ETX-90, have a diagonal mirror built in that can be flipped out of the way.

Figure 6.6 shows how diagonals work. Most star diagonals are 90° prisms and transmit about 90% of the light that reaches them. The losses occur by reflection where the light enters and exits the glass; the internal reflection is total.

The optical quality of prism diagonals varies. Total flatness is not critical because each point in the image reflects off only a small part of the diagonal surface. Local smoothness is more important and is generally good.

However, many cheap prism diagonals are miscollimated – they are not centered on the optical axis and not aligned at exactly 90°. To check for this problem, aim the telescope at a terrestrial object and make sure the same point remains centered as you turn the diagonal from side to side. Miscollimation is often easy to fix by taking the diagonal apart and putting it together again, making sure the prism fits into its mount properly.

Mirror diagonals of higher quality are available from Tele Vue and Lumicon. Their flatness meets the same rigorous standards as Newtonian secondaries.

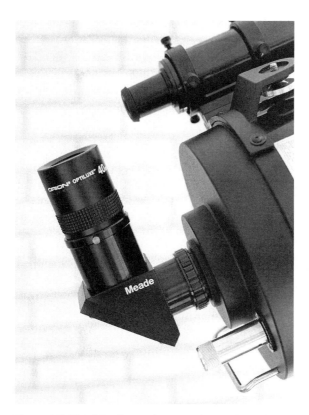

Figure 6.5. This Meade 2-inch mirror diagonal attaches directly to the back of the telescope. An adapter for $1\frac{1}{4}$-inch eyepieces is provided with it.

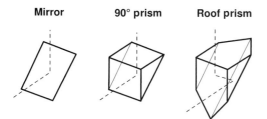

Figure 6.6. What's inside a star diagonal. Roof-prism type (right) gives an erect image.

Good mirror diagonals transmit slightly more light than prisms, typically about 95%. Also, unlike a prism, a mirror cannot introduce any chromatic aberration, even with low-$f$-ratio telescopes.

Despite these concerns, my own experience is that except for miscollimation, almost all the diagonals I have ever tried are good. Slight differences in light transmission are not perceptible, and optical quality, as judged by star tests, is not a problem. The limitation I most often run into is that a small diagonal may not completely illuminate the field of a low-power eyepiece.

A diagonal with a **roof prism** gives an erect image that is correct left to right. In place of the flat reflecting surface, it has a structure like a roof, with two surfaces at a 90° angle. As a result, the light is reflected twice instead of once and does not come out mirror-imaged. Each point in the image receives light from both surfaces, so the 90° angle is critical; even the slightest error results in blurring or doubling of the image. For that reason, roof prisms are not used for high-power work.

## 6.10.2 Barlow lenses

A Barlow lens (invented by Peter Barlow in 1833) is a negative (concave) lens that inserts into the telescope ahead of the eyepiece and enlarges the image (Figure 6.7). In effect, the Barlow lens increases the focal length and $f$-ratio of the telescope. In photography, a Barlow lens is known as a **teleconverter** and increases the focal length of a telephoto lens.

The Barlow lens does not change the range of powers that work well with a telescope, but it makes higher powers easier to use. Consider for example an 8-inch (20-cm) $f/10$ telescope. The highest usable power for such a telescope is about 400× and requires a 5-mm eyepiece (Table 6.1, p. 81). With a 2× Barlow, you can get 400× with a 10-mm eyepiece, which is much more likely to provide comfortable eye relief.

The gain in magnification depends on the distance from the Barlow lens to the eyepiece. Most 2× Barlows become 3× when inserted ahead of the diagonal instead of after it.

Single-element Barlow lenses are sometimes provided with small refractors, but good achromatic Barlow lenses have two or more elements. I use a Celestron Ultima 2× Barlow, which is a three-element apochromatic design. Tele Vue makes a four-element 5× Barlow that preserves parfocality with many Tele Vue eyepieces – unlike most Barlows, it does not require you to refocus after inserting it.

Higher-quality Barlow lenses are more critical for telescopes that have low $f$-ratios, since (just like eyepieces) they face a greater challenge intercepting a wider cone of light. The eyepiece after the Barlow, however, has a much easier

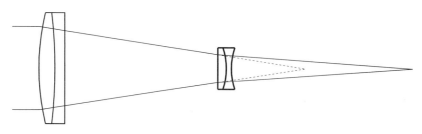

Figure 6.7. Barlow lens increases effective focal length and $f$-ratio, giving higher magnification.

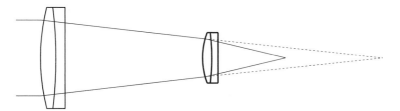

Figure 6.8. Focal reducer (compressor) reduces effective focal length and $f$-ratio, giving a smaller, brighter image.

Figure 6.9. Focal reducer attaches to telescope ahead of diagonal and eyepiece.

job. Inexpensive eyepieces that do not work well in an $f/4$ telescope may perform quite well when the telescope is converted to $f/8$ by a 2× Barlow – but the Barlow itself must be of top quality.

### 6.10.3 Focal reducers (compressors)

A **focal reducer** (**telecompressor** or just **compressor**) is the opposite of a Barlow lens – it is a positive (convex) lens that makes the image smaller and brighter, reducing the focal length and $f$-ratio (Figures 6.8, 6.9). Because it has to go

20 or 30 cm in front of the focal plane, a focal reducer can only be used with Schmidt–Cassegrains, Maksutov–Cassegrains, and some refractors.

Meade and Celestron make virtually identical four-element focal reducers that convert their $f/10$ Schmidt–Cassegrain telescopes to $f/6.3$ while reducing field curvature and other off-axis aberrations. Although marketed as a photographic accessory, this device can also be used visually and works well with a 26- or 32-mm eyepiece for rich-field viewing, at far lower cost and bulk than a 2-inch-diameter diagonal and eyepiece. I have used both configurations and frankly prefer the focal reducer.

The one drawback of a focal reducer is that the telescope can no longer fill as large a field. If you start with an image designed to cover 35-mm film and shrink it, it no longer covers 35-mm film. Some astrophotographers find this bothersome; I just accept it as inevitable.

### 6.10.4 Filters

#### Colored filters
Colored filters are used to bring out features on the planets and to dim the glare of the Moon. Table 6.2 gives specific examples.

Regardless of who makes them, these filters are identified by the equivalent Kodak Wratten filter numbers, and the threads are standardized so that filters of all brands fit nearly all eyepieces. Two-inch eyepieces take 48-mm camera lens filters.

If there is room, it is often more convenient to hold the filter by hand between your eye and the eyepiece.

Any filter darkens objects of the complementary color. For instance, the bluish-gray markings on Mars are brought out by a red filter. The belts of Jupiter, which are grayish-blue or purplish, stand out when viewed through a yellow filter.

A strongly colored filter also sharpens the image in unsteady air or when a planet is near the horizon. Like a prism, the air bends light of different

Table 6.2 *Filters for lunar and planetary observing*

| Color | Wratten numbers | Used for |
| --- | --- | --- |
| Red | 25 (dark), 23A (lighter) | Mars, some features on Jupiter and Saturn |
| Yellow | 15, 12 | Increased contrast on all planets, especially Jupiter and Saturn |
| Green | 58 (dark), 56 (medium) | Jupiter's Great Red Spot; polar caps on Mars; lunar surface detail |
| Pale blue | 80A, 82A | Bright comets; lunar surface detail |
| Blue | 38A (dark), 47 (very dark) | Mars during "blue clearings", Venus |

wavelengths differently, and by rejecting everything but a narrow band of wavelengths, the filter reduces the resultant blurring.

You can also get interesting results by using a nebula filter (next section) on the planets. Narrow-band nebula filters transmit a narrow segment of the spectrum in the blue-green range; broadband nebula filters block yellow while transmitting wavelengths above and below it. Both have been recommended particularly for viewing Jupiter.

### Light-pollution (nebula) filters

Nebula filters, also called light-pollution filters, darken the glare of city lights without filtering out stars and nebulae (Figure 6.10). This is possible because mercury-vapor and sodium-vapor streetlights emit light only at specific wavelengths. Thus, filters can be designed to block them out.

While blocking streetlights almost completely, such filters transmit much of the light of stars and galaxies, because starlight covers the whole spectrum. Hydrogen nebulae such as the Orion Nebula and Ring Nebula come out even better: they, too, emit only at narrow bands wavelengths, different from those of the streetlights, and their light can go through the filter almost unattenuated.

Filters of this type are expensive because they cannot be made of dyed glass. Instead, the glass is coated with several partly reflective layers whose thickness and spacing are tuned to specific wavelengths. The result is called an *interference filter* and looks almost like a mirror when viewed in daylight.

There are several kinds of light-pollution filters. The broadband type is best for viewing deep-sky objects other than hydrogen nebulae, for photography, and for use at mildly light-polluted sites; it cuts streetlight glare while transmitting as much of the spectrum as possible. Narrow-band (high-contrast) filters give the best view of hydrogen nebulae at heavily light-polluted sites; they transmit only the most visible wavelengths of ionized hydrogen, including hydrogen-beta and oxygen-III but not hydrogen-alpha.

Light-pollution filters do not help as much today as they did when first invented 20 years ago because a greater variety of outdoor lights is in use, and their emissions cover more of the spectrum.

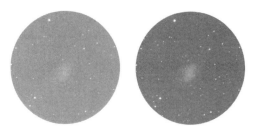

Figure 6.10. Nebula filter darkens the sky background, making nebulae more visible.

### Sun filters

Filters for viewing the Sun must be placed in front of the telescope, and only filters specially made for the purpose, and known to be safe, should be used. Loosely mounted aluminized Mylar filters give good results; high-quality glass solar filters are available at somewhat higher cost.

If not properly filtered, a telescope aimed at the Sun gathers enough light to blind the observer, melt its plastic components, and even set itself on fire. That is why the concentrated light of the Sun must never be allowed to enter the telescope. You cannot judge the safety of a solar filter by looking at it; most filter dyes transmit far too much infrared light – enough to cause serious eye injury – even if they look comfortably dark. For more information see *Celestial Objects for Modern Telescopes* and *Astrophotography for the Amateur.*

## 6.11  Eyeglasses

Do you need to wear glasses while observing? Quite possibly, for two reasons.

Some observers, including myself, prefer to wear glasses because we can't see the sky or our equipment without them. It would be easy enough to take my glasses off and focus the telescope to give a sharp image, but then I would have to put my glasses back on the moment I looked up. I prefer to keep them on all the time. That also enables me to focus the telescope so that other people can see through it.

Others have to wear glasses while observing because of astigmatism. The effect of astigmatism in the observer's eye is proportional to the exit pupil size, so if your eyes are only mildly astigmatic, you may get fine images at high powers, only to encounter distorted star images with your lowest-power eyepiece. Experiment to find out what the limitations are.

Eyeglasses vary in quality, and quality is worth paying for. The optical instrument that you use all the time is not the place to skimp. For years I have worn Zeiss Lantal lanthanum-glass lenses with a hard anti-reflection coating; the difference is noticeable.

In my opinion, all eyeglasses should be anti-reflection coated; otherwise you're throwing away 10% of the light. Glass lenses are of higher optical quality than plastic, though the difference is sometimes small; if you need plastic safety glasses for work or sports, consider getting a separate pair. If you wear bifocals, make sure you can comfortably look through the eyepiece with the distance-vision segment. (Seamless bifocals do not appeal to me; the transition between the two segments cannot give a perfectly sharp image at any distance.) Above all, make sure your eye doctor knows you're an amateur astronomer and that you need perfect vision at infinity focus in dim light.

In any case, *don't let your eyepieces scratch your glasses*, especially if you wear plastic lenses. Add-on rubber eyecups or small pieces of felt glued in place can protect your eyeglass lenses from this hazard.

## **6.12** Finders

Do computerized telescopes need finders? Yes, of course – for finding and centering the initial two stars, if nothing else.

Like binoculars, finders are rated by magnification and aperture. Thus, a $5 \times 24$ finder magnifies $5\times$ and has an aperture of 24 mm. Larger finders, such as $8 \times 50$ or $10 \times 60$, provide interesting views of star clusters in their own right and are precise enough to center objects for CCD imaging.

Should the finder have a diagonal? Maybe. An "elbow" or right-angle finder is much easier to use when sighting Polaris during polar alignment (Figure 4.7, p. 46). Particularly on smaller telescopes, access to a straight-through finder in that situation is blocked by the base of the telescope.

But a straight-through finder (Figure 6.11) is much easier to use the rest of the time. My technique is to keep both eyes open, one of them looking through the finder and the other viewing the star directly, then move the telescope to make the two images coincide.

With a computerized telescope, you do not need to see faint stars through the finder. Accordingly, nonmagnifying LED finders (Figure 6.12) are popular and convenient. Using a beamsplitter, an LED finder displays a red dot that seems to be superimposed on the sky. You simply sight along it, looking directly at the sky, and put the spot on the star that you want to view. A more elaborate type, the Telrad (invented by Steve Kufeld and available through many telescope dealers), projects a pattern of circles.

Sometimes you need both kinds of finders on the same telescope. An ETX-90, for instance, is much easier to polar-align if you have both a straight-through

Figure 6.11. Straight-through $8 \times 50$ finder on a Meade LX200.

Figure 6.12. Nonmagnifying LED finder on Celestron NexStar 5. Seen through its window, a red spot appears superimposed on the sky.

LED finder for rough alignment and a right-angle magnifying finder for precise pointing. Fortunately, LED finders are easy to add; they attach with adhesive pads, requiring no modification to the telescope.

# Chapter 7
# Astrophotography

## 7.1 Overview

Photography gives you a way to record what you see through the telescope. Surprisingly, though, astrophotography does not reproduce what you see visually. In lunar and planetary work, it is very hard to get pictures as sharp as what the eye can see, because the eye can seize moments of atmospheric steadiness in a way that the camera cannot. In deep-sky work, on the other hand, the camera often records far more than the eye could see with the same instrument because film can accumulate light in a long exposure.

This chapter will tell you enough about astrophotography to get you started. It is not a complete guide; for that, see my other book, *Astrophotography for the Amateur* (Cambridge University Press, 1999).

One word of advice: astrophotography is a matter of skill, not just equipment. It is definitely not a matter of "You press the button, we do the rest" – it is a stiff test of how well you understand your equipment and the principles on which it operates. Never buy a piece of equipment until you know exactly what you'll use it for.

There is also an element of luck. The pictures that you see published in magazines are the work of experienced astrophotographers and are generally the best of many, many tries. Do not expect to equal them immediately.

But some techniques do yield very good pictures even in the hands of a beginner. Piggybacking is one example; so is afocal photography of the Moon (Figure 7.1). It is rewarding to share your pictures with local groups, enter them in art exhibits, and so forth, even if they can't compete with those of the world's top astrophotographers. The beauty of a picture is not proportional to the difficulty of taking it or the cost of the equipment. After years of experience, I keep coming back to the most basic techniques because they are the most satisfying.

In what follows I'm going to assume that you already understand basic photography – how a lens forms an image, how shutters and $f$-stops control exposure, and how film is developed and printed. If not, some reading is in order.

Figure 7.1. Pictures like this are within easy reach of the beginning astrophotographer. Olympus 2.1-megapixel digital camera, handheld, aimed into eyepiece of 8-inch $f/10$ Schmidt–Cassegrain telescope with 26-mm Plössl eyepiece.

One of the best guides is *Basic Photography*, by Michael Langford (Butterworth-Heinemann, 2001) and, on a more technical level, *The Camera*, by Ansel Adams (Bulfinch Press, 1987).

## 7.2 Attaching cameras to telescopes

### 7.2.1 Optical configurations

Figure 7.2 shows six ways to attach a camera to a telescope. One of them, **piggybacking**, does not take pictures *through* the telescope at all; instead, the camera

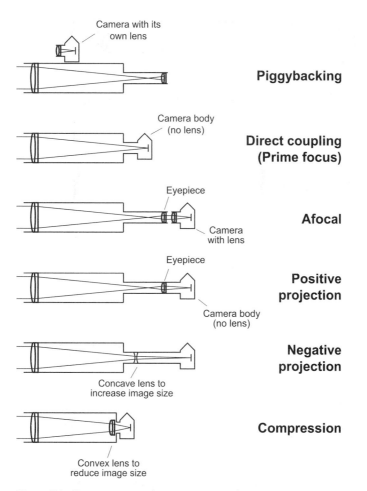

Camera with its own lens

**Piggybacking**

Camera body (no lens)

**Direct coupling (Prime focus)**

Eyepiece

**Afocal**

Camera with lens

Eyepiece

**Positive projection**

Camera body (no lens)

**Negative projection**

Concave lens to increase image size

**Compression**

Convex lens to reduce image size

Figure 7.2. Six ways to attach a camera to a telescope.

takes a long exposure through its own lens while the telescope tracks the stars. The other five methods use the telescope as a giant telephoto lens.

The **direct** or **prime focus** method is the simplest. The telescope, without an eyepiece, goes onto the camera, without a lens. Most refractors, Schmidt–Cassegrains, and Maksutov–Cassegrains work well in this configuration, but many Newtonians will not reach focus because the image plane is not far enough from the end of the eyepiece tube.[1]

The **afocal** method works with any telescope and any camera. The camera, with its lens, is aimed into the eyepiece of the telescope. This is a simple and

[1] The term *prime focus* is potentially confusing because, on large observatory telescopes, it refers to the focus of the main mirror, as opposed to the Newtonian or Cassegrain focus. Now that some amateur telescopes, such as Celestron's Fastar system, can be used this way, I prefer the term *direct coupling* or *direct method*. I thank Lenny Abbey for pointing out the potential for confusion.

foolproof system, especially with digital and video cameras, but I also use it often with 35-mm SLRs.

In afocal coupling, the camera lens is always wide open (at its lowest-numbered $f$-stop) and focused on infinity. Digital and other autofocus cameras should be locked on infinity focus. The camera can be held in place with a bracket or can stand on its own tripod. When photographing the Moon with a short exposure, you can even handhold the camera.

**Positive projection (eyepiece projection)** is one way to get greater magnification than with the direct method. The telescope eyepiece is in place, but the camera, behind it, has no lens. Although popular with amateurs, this method, in my experience, does not give very good optical quality; images are usually sharp only at the center, a fact that is often obvious in images of the Moon.

Like afocal coupling and unlike the direct method, positive projection works well even with Newtonians and other telescopes that cannot put the image plane very far past the end of the eyepiece tube.

**Negative projection** is like the direct method with a Barlow lens (p. 92) or teleconverter added for extra magnification. (A teleconverter is, after all, nothing but a Barlow lens for a camera.) I find it much better than positive projection for imaging the Moon and planets.

**Compression** is simply the direct method with a focal reducer added to reduce the magnification and brighten the image (p. 93). It is often used in deep-sky work.

## 7.2.2 Brackets and adapters

A "camera adapter" for a telescope can be any of a wide variety of gadgets, depending on the kind of camera, the kind of telescope, and the way you want to couple them.

Piggybacking is simplest because all you have to do is mount the camera on the outside of the telescope. Larger telescopes have predrilled holes for piggyback brackets. With smaller telescopes, including the NexStar 5 and ETX-90, you have to attach the camera with a strap, easily made from a drainpipe hanger or a large hose clamp. You can even mount a camera on the counterweight of a German equatorial mount.

There are three ways to do afocal coupling. You can support the camera on a separate tripod – which looks clumsy but is marvelously free of shutter vibration. Or you can use an afocal adapter. This in turn may be either a bracket that supports the camera by its tripod socket, or a cylindrical device that grips the eyepiece and screws into the front of the camera lens like a filter. Afocal adapters are available from ScopeTronix (1423 S.E. 10th Street, Unit 1A, Cape Coral, FL 33990, http://www.scopetronix.com) and LensPlus (11969 Livona Lane, Redding, CA 96003, U.S.A., http://www.lensadapter.com) among others.

Figure 7.3 shows how direct (prime focus) coupling is done. The camera body accepts a T-adapter (T-ring), which has $42 \times 0.75$-mm threads that accept other

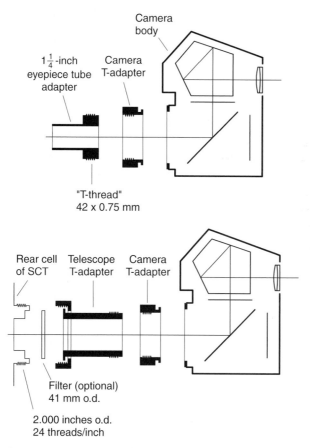

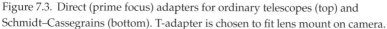

Figure 7.3. Direct (prime focus) adapters for ordinary telescopes (top) and Schmidt–Cassegrains (bottom). T-adapter is chosen to fit lens mount on camera.

accessories. (These T-adapters are all that is left of a universal lens mount system that was popular in the 1960s but died out when lenses began to require aperture coupling.) Negative projection is done with the same adapter, with a Barlow lens between the adapter and the telescope or a teleconverter between the T-ring and the camera.

Adapters for eyepiece (positive) projection are similar but larger, with space inside for the eyepiece. They do not fit large wide-angle eyepieces.

## 7.3 Two simple projects to get you started

### 7.3.1 Project #1: the Moon, afocal method

If you are eager to get started, here is a simple project you can try right away with almost any telescope.

Load your camera with any general-purpose print or slide film. Aim the camera into the eyepiece of your telescope, either by using a bracket to hold it in place or by putting it on its own tripod. Set the camera lens to infinity focus and widest aperture (lowest $f$-stop). Get the Moon in view, focus the telescope carefully, and take a picture. If your camera has through-the-lens auto exposure, it will probably auto-expose correctly; otherwise, try a wide range of exposures, such as $\frac{1}{30}$ to $\frac{1}{250}$ second. Figures 7.1 and 7.4 show what to expect.

Figure 7.4. Even a cheap camera can take good afocal pictures. Russian Lubitel twin-lens reflex camera, on separate tripod, aimed into the eyepiece of a 5-inch (12.5-cm) Schmidt–Cassegrain telescope at 40×. Exposure $\frac{1}{30}$ second on Ilford HP5 Plus film.

This technique works with all types of cameras, whether SLR, digital, or video. You can even use cameras that do not let you focus through the lens. In that case, you will need a small hand-held telescope or monocular; your finder or half of a pair of binoculars will do. Focus the hand-held telescope on the Moon or stars, then aim it into the eyepiece of the main telescope. Focus the main telescope so that what you see is sharp. Voilà – you've set the main telescope to produce a virtual image at infinity. Now put the camera in place, with its lens set to infinity, and you'll get a sharp image.

### 7.3.2 Project #2: the stars, piggybacking

Taking dramatic pictures of the stars and Milky Way is easy if you have an equatorially mounted telescope (Chapter 4). Load your camera with color slide film that has low reciprocity failure and a strong response to deep red light; Kodak Elite Chrome 200 or E200 Professional is an excellent choice. Attach the camera to the telescope somehow, either on a piggyback bracket or by any convenient means. The camera and telescope need not point in exactly the same direction, though they should be reasonably close.

On a clear moonless night, take a 2- to 5-minute exposure of the sky with the camera lens wide open, while the telescope tracks the stars. With a 100-mm or shorter lens, you will probably not have to make guiding corrections. Use shorter exposures under a bright town sky, longer exposures in the country. Figures 7.5 and 7.6 show what you can achieve.

Figure 7.5. Dramatic view of the Milky Way in Sagittarius. No guiding corrections were needed during this 2-minute piggyback exposure on Kodak Elite Chrome 200 slide film with a 50-mm lens at $f/2.8$.

Figure 7.6. The Belt of Orion and Orion Nebula. One-minute piggyback exposure (without guiding corrections) on Elite Chrome 200 with a 90-mm lens at $f$ /2.8, under a town sky. In the country, a longer exposure would bring out far more nebulosity.

If, on the finished slides, the stars are elongated north to south, your polar alignment wasn't accurate enough; if they are elongated east to west, your tracking wasn't smooth enough. Either way, guiding corrections during the exposure will alleviate the problem.

Can you do anything like this on an altazimuth mount? Maybe, if your tracking motors are fairly smooth. The trick is to choose a star field fairly low in the eastern or western sky, where field rotation is at a minimum (p. 39), and expose for no more than two or three minutes. That should be enough to whet your appetite for an equatorial wedge.

Why do I recommend slide film? So that you won't be at the mercy of the people or machines that make the prints. Most photo labs do not know how to print negatives of star fields; the pictures come out either too light or too dark. With slides, you can see what the camera actually recorded, and you can then order custom prints that match the slides.

## 7.4 Equipment for astrophotography

### 7.4.1 Telescope requirements

You can do lunar and planetary photography with almost any telescope as long as the mount is steady. Deep-sky work requires an equatorial mount (or a fork mount on an equatorial wedge) to prevent field rotation.

Schmidt–Cassegrain and Maksutov–Cassegrains are the easiest to couple to cameras because you can easily put the focal plane where you want it, right at the eyepiece or deep inside a camera body. Refractors are almost as easy to work with. As already noted, Newtonians are less versatile than other types because their focal plane cannot be placed very far beyond the end of the eyepiece tube; that means compression and direct coupling are likely to be impractical, but afocal and projection photography are no problem.

When coupled directly to a camera body, very few telescopes fill the field of a 35-mm camera; after all, they are designed to work with eyepieces a good bit smaller than 35-mm film. Two-inch eyepiece tubes help somewhat, but there is usually still some vignetting.

The 10-inch and larger Meade LX200s, and the 11-inch and larger Celestrons, are exceptions. These telescopes have a back that matches that of the 8-inch, but the central part of it can be removed to give a larger opening. Camera adapters that make use of the larger opening are available from Lumicon (2111 Research Drive, Livermore, CA 94550, U.S.A., http://www.lumicon.com) and other suppliers; they produce images that are free of vignetting in direct mode and only slightly vignetted when a compressor is used.

Smooth tracking and a steady mount are vital for deep-sky work. Many of the less expensive computerized telescopes, such as the Meade ETX-90, are not designed for photography and are usable only with difficulty. Others, such as the Meade LX200 and Celestron Ultima 2000, are fine photographic instruments. For smooth tracking, you need a well-made worm-gear drive, preferably with periodic-error correction (PEC, p. 53) so that the irregularities in the gears can be memorized by the computer and counteracted automatically.

### 7.4.2 35-mm SLR cameras

Most amateur astrophotography is done with 35-mm single-lens reflex (SLR) cameras. As Figure 7.7 shows, an SLR has a built-in mirror and focusing screen so that you can see the actual image formed by the lens or telescope. The mirror flips up before the shutter, behind it, opens to make an exposure.

For astrophotography, older SLRs are better than the most modern ones. New features such as autofocus and auto exposure are not needed. Instead, what you need is full manual control, combined, if possible, with the ability to make long exposures without running down the batteries (though a supply of extra

Figure 7.7. Cross-section of a 35-mm SLR. The mirror intercepts light for focusing on the screen, then flips out of the way before the shutter opens. (Olympus, reproduced by permission.)

batteries may be cheaper than a new camera). The ability to change focusing screens is useful; if the screen is not interchangeable, you will have to make do by focusing on the smoothest part of it, not the central prism or split-image device.

To keep the camera from shaking when the mirror flips up, it is handy to have either **mirror lock** or **mirror prefire**. Mirror lock means there is a separate button to flip the mirror up in advance of the exposure. Mirror prefire means that when you make a delayed exposure with the self-timer, the mirror flips up at the beginning of the cycle, several seconds before the shutter opens.

People often ask me which SLR is best for astrophotography. The truth is that almost any SLR can be used to some extent, so if you already have a camera, see what you can do with it. Also look for older cameras (vintage about 1970) that may have been languishing in your relatives' or friends' closets; almost all SLRs from the 1970s are good choices. For a chart of many suitable camera models and their features, see *Astrophotography for the Amateur*. But before shopping for a new camera, do as much as possible with whatever you already have.

### 7.4.3 Other film cameras

Some amateurs do astrophotography, particularly piggybacking, with medium-format SLRs such as the Hasselblad, Bronica, Mamiya, and Pentax. The lack of grain in larger-size negatives is helpful.

For piggybacking and afocal coupling, the camera need not be an SLR. It is sufficient that you be able to focus it to infinity and control the exposure manually. For afocal photography, the telescope can be focused with a separate hand-held telescope (p. 105). I have done both piggybacking and afocal photography with a scale-focusing Voigtländer Vito B that is more than 50 years old.

Unfortunately, most automated "point-and-shoot" cameras are completely unsuitable. The exposure cannot be controlled manually, and the lens is, in any case, tiny – often $f/9$ or slower. If you want a versatile camera at a low price, look for a fully adjustable one from the 1970s or earlier.

### 7.4.4 Digital and video cameras

You can get good images of the Moon and planets with an ordinary digital or video camera aimed into the eyepiece of the telescope. It is easy to focus and adjust the exposure because you can see immediately what you are getting. The camera focus should be locked at infinity so that all focusing can be done with the telescope.

Some digital cameras give you manual control of exposure. Others let you adjust exposure indirectly with a "lighten/darken" adjustment. In some cases, the camera may insist on overexposing a small planet image seen against a dark background; in that case your only resort may be to add a neutral density filter between the eyepiece and the camera.

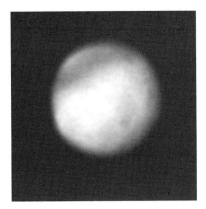

Figure 7.8. Mars. Olympus 2.1-megapixel digital camera aimed into 9-mm eyepiece on an 8-inch Meade LX200 (222×). This is the best of several images, and it was enhanced with Adobe *Photoshop LE*.

Digital cameras and camcorders are not suitable for deep-sky work. The reason is that their CCDs (charge-coupled device image sensors) are not cooled, and in exposures longer than about one second, there is excessive noise, which shows up as a speckled pattern.

### 7.4.5 Astronomical CCD cameras

CCD cameras designed specifically for astronomy use thermoelectric coolers to keep the image sensor at a low temperature, where it is less subject to noise (random leakage of electrons), so that long exposures are possible (Figure 7.9). These cameras perform impressively on both planets and deep-sky objects, and because the CCD's response to light is perfectly linear, it is relatively easy to subtract out the effects of city lights, making it possible to image faint nebulae and galaxies even in the suburbs.

CCD cameras for amateur use are made by Meade; SBIG (Santa Barbara Instrument Group, 147-A Castilian Drive, Santa Barbara, CA 93117, U.S.A., http://www.sbig.com); Apogee (11760 Atwood Road, #4, Auburn, CA 95603 U.S.A., http://www.ccd.com); Starlight Xpress (Ascot Road, Holyport, Berkshire, SL6 3LA, U.K., http://www.starlight-xpress.co.uk); and other companies. CCD cameras generally have to be connected to a laptop computer during use, but the SBIG STV includes its own controller and video screen.

Professional astronomers switched from film and plates to CCDs in the 1980s, and amateur astrophotographers are following suit. Unfortunately, CCDs within reach of amateur budgets give rather small images, often as small as $320 \times 200$ pixels; if you like big prints, you should (at least at present) stick to film. Soon, however, $1000 \times 1000$-pixel or $2000 \times 2000$-pixel imaging will be within reach of amateur budgets, and then film will face serious competition from CCDs.

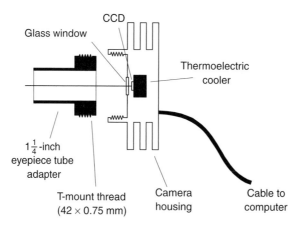

Figure 7.9. An astronomical CCD camera includes a thermoelectric cooler to permit long exposures. (From *Astrophotography for the Amateur*.)

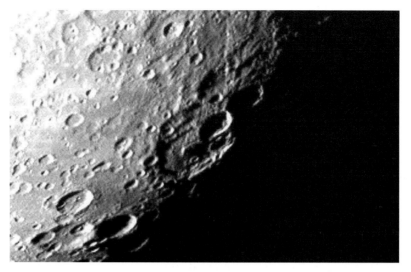

Figure 7.10. CCD image of the walled plain Janssen on the Moon, taken with an SBIG STV coupled directly to an 8-inch (20-cm) Meade LX200 telescope. The air was generally unsteady, but the camera was able to make the most of a moment of steadiness.

One important advantage of CCD cameras for lunar and planetary work is that it is easy to take large numbers of images and keep only the best. Some cameras can even select the best images automatically during the observing session. This makes it possible to seize brief moments of atmospheric steadiness just as an experienced visual observer does (Figure 7.10).

CCD cameras invariably plug into the telescope in place of the eyepiece; a Barlow lens or a compressor can be used ahead of the camera to change the image size. Almost all CCD cameras can also be used as autoguiders (p. 118).

## 7.5 Focal length, image size, and $f$-ratio

### 7.5.1 Finding the effective focal length

Just how much magnification does your setup give you? In a sense, the "magnification" of a picture is almost meaningless because any picture that you take of a planet will be far smaller than the planet itself. Instead, what you want to know is the field of view or the image size. To find that, you must first find the effective focal length (EFL):

- With an ordinary camera lens (as in piggybacking), the EFL is simply the focal length.
- With direct (prime focus) coupling, the EFL is the focal length of the telescope. For example, the focal length of a Meade ETX-90 is 1250 mm. (Throughout this book, I give all focal lengths in millimeters, no matter how large.)
- With afocal coupling:

    EFL = Telescope magnification × Camera lens focal length

For example, a camera with a 50-mm lens looking into a 100× telescope has an EFL of 5000 mm. With digital and video cameras, the focal length of the lens is often unknown, but the field of view is easy to determine by direct experimentation.

- With positive projection:

$$\text{EFL} = \text{Telescope focal length} \times \frac{S - F}{F}$$

where $F$ is the eyepiece focal length and $S$ is the distance from the eyepiece to the film. For example, with a Meade ETX-90 (focal length 1250 mm) and a 25-mm eyepiece located 125 mm from the film, the EFL is $1250 \times (125 - 25)/25 = 1250 \times 4 = 6000$ mm.

- With negative projection, if you're using a teleconverter, you can rely on it to give exactly its rated magnification. Thus a telescope of 1250 mm focal length and a 2× teleconverter will have an EFL of $1250 \times 2 = 2500$ mm.

    With a Barlow lens, the best thing to do is experiment, since you normally do not know the focal length of the Barlow lens. As a rule of thumb, most Barlows give 50% to 100% more magnification when used ahead of a camera than when used ahead of an eyepiece, since the image plane is further from the Barlow lens. For example, a 2× Barlow might multiply the telescope focal length by 3 or 4. To get a more exact value, photograph the Moon and measure the size of the image.

- With compression, the focal length obtained with a particular compressor lens is generally specified by the manufacturer. For example, Meade's $f/6.3$ compressor for $f/10$ telescopes multiplies the telescope focal length by 0.63.

### 7.5.2 Image size and field of view

The size of the image on the film depends on the apparent size of the celestial object and the EFL, thus:

$$\text{Image size (mm)} = \frac{\text{Apparent size of celestial object (arc-seconds)}}{206\,265} \times \text{EFL (mm)}$$

For example, the apparent diameter of Jupiter is typically 40″, so with an EFL of 5000 mm, you get images of Jupiter that are $(40/206\,265) \times 5000 = 0.9$ mm in diameter. It follows that you need an EFL of at least 5000 mm to get usable images of the planets, and even then, the images are small.

There is a related formula for finding the field of view. Let's assume that the usable area of a piece of 35-mm film is about 19 mm in diameter; that is reasonable because there is usually some vignetting and edge blurring. Then the formula simplifies to:

$$\text{Field of view (arc-seconds)} = \frac{4\,000\,000}{\text{EFL (mm)}}$$

For example, with an EFL of 1250 mm, the field of view is 3200″, and the Moon, which is 1800″ in diameter, fits comfortably into the picture. As a rule of thumb, use EFLs between 1000 and 2000 mm to photograph the full face of the Moon; use longer EFLs to enlarge selected areas.

### 7.5.3 Finding the *f*-ratio

To calculate exposures, you need to know the *f*-ratio of the complete system. Fortunately, that's easy:

$$f\text{-ratio} = \frac{\text{Focal length (mm)}}{\text{Telescope aperture (mm)}}$$

Note that the focal length and aperture must be given in the same units.

For example, consider again the Meade ETX-90 with eyepiece projection giving an EFL of 6000 mm. The aperture is 90 mm, so the *f*-ratio is 6000/90 = 66.7. That is, you have turned a telescope into a 6000-mm *f*/66.7 telephoto lens. High *f*-ratios, *f*/45 and higher, are common in lunar and planetary work.

### 7.5.4 Exposure, film, and development

In astrophotography, you have to buy your film by name, not just by speed. For example, Kodak Elite Chrome 200 and Kodachrome 200 are both 200-speed color slide films, but they are as different as night and day. Kodachrome 200 has severe **reciprocity failure**, which means that it loses speed at low light levels, in long exposures. In a ten-minute exposure it is a great deal slower than Elite Chrome 200, which has very little reciprocity failure.

The fastest films are not the best, for several reasons. Faster films generally have more reciprocity failure, so Elite Chrome 400 is actually worse than Elite Chrome 200 in long exposures. Faster films also have more grain, which is undesirable since it hides star images and planetary detail.

Films also differ in their response to the hydrogen-alpha wavelength, 656 nm, in the deep red. That is the main emission from the Lagoon Nebula, North America Nebula, and their kin. On some films, such as Elite Chrome 200, these nebulae are bright red; on other films, such as Tri-X Pan, they hardly show up at all. The Kodak Elite Chrome films have a strong response to hydrogen-alpha; so do the Kodak Supra (not Portra) color negative films. Fuji color films have a weaker response to hydrogen-alpha, and most black-and-white films have no response at all.

One exception is Kodak Technical Pan film, which responds strongly to hydrogen-alpha but requires treatment with hydrogen gas ("hypering") to increase its speed and reduce reciprocity failure. Hypered Technical Pan was a mainstay of astrophotography in the 1980s; today, the best color films pick up nebulae almost as well and are much easier to work with.

Table 7.1 *Exposure table for Kodak Elite Chrome 200 (E200) film, roughly correct for other 200- and 400-speed films*

| Object | 2.8 | 4 | 5.6, 6.3 | 8 | 10, 11 | 16 | 32 | 64 | 128 |
|---|---|---|---|---|---|---|---|---|---|
| | | | | *f*-ratio | | | | | |
| Moon (thin crescent) | 1/125 | 1/60 | 1/30 | 1/15 | 1/8 | 1/4 | 1 sec | 4 sec | — |
| Moon (half) | 1/500 | 1/250 | 1/125 | 1/60 | 1/30 | 1/15 | 1/4 | 1 sec | 4 sec |
| Moon (full) | — | 1/2000 | 1/1000 | 1/500 | 1/250 | 1/125 | 1/30 | 1/8 | 1/2 |
| Moon (partial eclipse) | 1/125 | 1/60 | 1/30 | 1/15 | 1/8 | 1/4 | 1 sec | 4 sec | — |
| Moon (total eclipse) | 1 sec | 2 sec | 5 sec | 10 sec | 20 sec | — | — | — | — |
| Sun (through Baader visual filter) | — | — | — | 1/2000 | 1/1000 | 1/500 | 1/125 | 1/30 | 1/8 |
| Sun (total eclipse, no filter) | 1/60 | 1/30 | 1/15 | 1/8 | 1/4 | 1/2 | — | — | — |
| Mercury | — | — | — | 1/125 | 1/60 | 1/30 | 1/4 | 1 sec | 4 sec |
| Venus | — | — | — | 1/2000 | 1/1000 | 1/500 | 1/125 | 1/30 | 1/4 |
| Mars | — | — | — | 1/250 | 1/125 | 1/60 | 1/8 | 1/2 | 2 sec |
| Jupiter | — | — | — | 1/125 | 1/60 | 1/30 | 1/4 | 1 sec | 4 sec |
| Saturn | — | — | — | 1/15 | 1/8 | 1/4 | 1 sec | 4 sec | — |
| Comets, bright nebulae (typical) | 2 min | 4 min | 8 min | 15 min | 30 min | — | — | — | — |
| Galaxies, faint nebulae (typical) | 6 min | 12 min | 24 min | 1 hr | — | — | — | — | — |

Table 7.1 gives exposures for a variety of celestial objects on Elite Chrome 200. The table is approximately valid for most other 200- and 400-speed films. When in doubt, vary your exposures, using the table only as a rough guide. There is no specific "correct" exposure for most astronomical photographs; the best exposure depends on what you want the picture to look like. For more exposure tables and explanations of how exposures are calculated, see *Astrophotography for the Amateur.*

## 7.6 Focusing and sharpness

Good lunar and planetary photography requires trying the same thing over and over until you finally get a picture in which the air is steady at the right moment. It also requires accurate focusing and freedom from vibration.

Consider focusing first. A camera attached to a telescope is much harder to focus than the same camera with an $f/1.8$ lens. The image does not snap into focus at any particular point. Instead, finding the point of best focus requires careful attention. The focus knob may have backlash; that is, the correct position may depend on whether you were last turning it clockwise or anticlockwise. The image through the telescope at a long EFL is sure to be somewhat blurry no matter how carefully you focus.

The central split-image or microprism area in an SLR focusing screen is useless with telescopes. Instead, focus on the smoothest matte area of the screen, typically a ring surrounding the central spot. If at all possible, change to a focusing screen that has a fine matte surface all over, such as a Nikon B screen or a Beattie Intenscreen.

Other types of focusing screens, such as clear screens with crosshairs, are popular with advanced astrophotographers, but their usage is tricky. Remember that the purpose of a focusing screen is not to make the image look sharp; it is to make it look blurred when it is out of focus. Images on clear-crosshairs screens tend to look like they are in focus when they're not.

If you find focusing difficult, check whether you can see the focusing screen clearly. Perhaps the camera eyepiece is not in focus for your eyes; a corrective lens can be added, or you can use an adjustable magnifier. The Olympus Varimagni Finder fits not only Olympuses, but also most Minoltas, Pentaxes, and Yashicas, among others (but no Nikons).

Finally, remember that not all blurriness is caused by incorrect focusing. Shutter vibration is a *serious* challenge to the lunar and planetary photographer. I usually get around it by choosing a film and $f$-ratio so that the exposure is $\frac{1}{4}$ second or more, and then doing a **hat trick**. That means holding my hat or a large black card in front the telescope; opening the shutter; then carefully moving the hat or card away and back again. The hat or card functions as a vibration-free shutter. In deep-sky work, shutter vibration is less of a problem because when the exposure lasts several minutes, a millisecond or two of vibration is only an infinitesimal part of it.

## 7.7 Deep-sky techniques

Deep-sky photography is at once easier and harder than lunar and planetary work. It's easier because you have a better chance of getting a good picture on the first try. It's harder because additional equipment and skills are needed.

As already noted, you need an equatorial mount or wedge. You also need an eyepiece with crosshairs for guiding. No matter how good your mount is, the drive motors in it are not perfect, and you will need to make manual corrections in the east–west direction to smooth out the motion. Corrections in the north–south direction will counteract small errors in polar alignment.

Figure 7.11. The Orion Nebula (M42). Five-minute exposure with an 8-inch Meade LX200 and a compressor giving effective $f/5.6$, autoguided with an SBIG STV CCD camera at the eyepiece position of an off-axis guider. Kodak Elite Chrome 200 film pushed one stop.

Figure 7.12. The North America Nebula and other nebulosities in Cygnus. Twenty-minute piggyback exposure on Ektachrome Elite II 100 film pushed two stops, using a 90-mm lens at $f/2.8$ and a broadband nebula filter. The slide was scanned and processed digitally to increase contrast.

For piggybacking, guiding is easy: mount the camera on top of the telescope and watch a star through the telescope itself, keeping it on or near the crosshairs. Guiding tolerances are discussed in detail in *Astrophotography for the Amateur*; suffice it to say that when piggybacking with a medium telephoto lens, you do not have to keep the star precisely on the crosshairs, merely near them.

I have had good results piggybacking with a NexStar 5, a lightweight telescope not designed for photography. The secret to good guiding is to make sure the tracking is set to "EQ North" so that only one motor is running; set the slewing rate to the lowest value; and turn off backlash compensation so there will be no sudden jerks. It is better for guiding corrections to be delayed than for them to be too sudden or irregular. On more advanced telescopes such as the Meade LX200, guiding goes very smoothly.

When you are photographing through the telescope, guiding is more of a challenge. For one thing, there is no room for error; the star must stay *on* the crosshairs. But the bigger question is, if you are using the main telescope for photography, how do you guide?

One solution is to use a separate guidescope. This tactic works well with refractors and reflectors, provided the guidescope has a high enough magnification

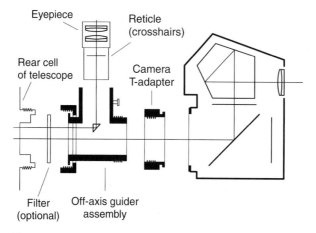

Figure 7.13. An off-axis guider intercepts a portion of the main image that would not fall on the film. Finding a suitable guide star is often difficult.

(200× is sufficient; 500× is better, though it's far too high for real observing). A Barlow lens ahead of the guiding eyepiece helps improve precision.

Separate guidescopes do not work so well with Schmidt–Cassegrains and Maksutov–Cassegrains because the mirror tends to shift slowly as the telescope tilts while following a star. Thus the image in the telescope undergoes a movement that the guidescope does not see.

The solution is to use an **off-axis guider** (sometimes abbreviated OAG), intercepting a small part of the image that would not fall on the film, so that you can guide on the same image that the camera is photographing (Figure 7.13). The hard part about using an OAG is finding a suitable guide star; all too often it seems that there is nothing brighter than twelfth magnitude in the right place. The more ways you can adjust the OAG to change the view, the better. Finding guide stars is much easier when you are photographing star clusters and nebulae in rich portions of the Milky Way than when you are photographing galaxies.

Guiding is tedious, but, fortunately, most CCD cameras can do it for you; a CCD camera used for this purpose is called an **autoguider**. Meade makes a low-end CCD camera, the 201XT, that can only guide on relatively bright stars; it's small and inexpensive. I use an STV from SBIG (the Santa Barbara Instrument Group), which autoguides very well and reports its accuracy as it goes. Either of these goes in place of the crosshairs eyepiece that you would otherwise use. An exception is the Apogee Lisää, whose housing can be set up to work as an off-axis guider assembly in front of a 35-mm camera or another CCD.

But when you are taking a CCD image, how do you guide? Unfortunately, the same CCD cannot make guiding corrections and expose an image at the same time. One solution is to use a second CCD in an off-axis guider – a time-honored but expensive practice. More recently, several vendors have found ways to put a single CCD to two uses. Starlight Xpress cameras can track and record at the

Figure 7.14. 320 × 200-pixel image of the globular cluster M13. SBIG STV CCD camera, 8-inch telescope with $f/5.6$ compressor, combination of three 15-second exposures in track-and-accumulate mode.

same time, using alternate rows of pixels on the same CCD. The SBIG STV can "track and accumulate", which means that it makes a short exposure, checks for image shift, makes another short exposure, shifts it as needed to match, and adds the two together, over and over.

## 7.8 Digital image processing

The CCD revolution is almost insignificant compared with the digital image processing revolution. Whether they originated on film or on CCD, astronomical images can now be processed by computer to enhance contrast, correct color, and bring out faint details. Although there is plenty of special software for the purpose, such as Software Bisque's *CCDSOFT*, you can also do very good work with Adobe *Photoshop Elements* (formerly *Photoshop LE*) and similar products.

The most important parameter of a digital image is the number of pixels. As a rule of thumb, a 200 × 300-pixel image looks good when filling a small part of a computer screen. For a whole computer screen or a sharp postcard-sized image, you need at least 900 × 1200 pixels. For large prints, you need even more.

Most CCD images are on the order of 200 × 300 or 400 × 600 pixels. Thus, one of the first steps in processing them is to **resample** (enlarge) the image to increase the number of pixels. This does not bring out additional detail, but it does decrease the stairstep or boxlike appearance of fine detail in the image.

Film images, on the other hand, already have plenty of pixels. A 35-mm slide or negative, scanned with a good film scanner, can easily comprise 2400 × 3600 pixels (8.6 million pixels, 8.6 **megapixels**). It is often better to scan at half that resolution, producing a $1200 \times 1800$-pixel (2.1-megapixel) image that does not show film grain.

Film should always be scanned on a film scanner, not on a flatbed scanner with an adapter. Although some flatbed scanners are now *almost* good enough, the first generation of them did not have nearly enough resolution to do justice to film. Flatbed scanners usually have a true resolution of 300 to 600 pixels per inch (12 to 14 pixels/mm). (*Interpolated* resolution doesn't count; it is created by resampling the image and does not pick up additional detail.) Film scanners resolve 2400 pixels per inch (90 pixels/mm) or more and pick up all the detail that the film records. A film scanner also needs enough **dynamic range** (brightness range) to capture both the brightest and the darkest areas of the picture. Negatives have less dynamic range than slides and are easier to scan.

Once you have the image in the computer, what do you do with it? Figures 7.15 and 7.16 show two basic operations. You can sharpen the image by **unsharp masking** – that is, by exaggerating the differences between adjacent pixels. Originally an "unsharp mask" was a blurred negative, sandwiched with a sharp positive to smooth out large gradients without hiding fine detail; today unsharp masking is done by averaging and subtraction. It brings out detail in a dramatic way, particularly on lunar images. It can also bring out film grain.

Figure 7.15. Digital image enhancement at work. *Left:* Original image, taken on Fuji Sensia 100 color slide film with a 5-inch (12.5-cm) $f/10$ Schmidt–Cassegrain, then scanned and converted to monochrome. *Right:* Same, after unsharp masking and Gaussian sharpening. (From *Astrophotography for the Amateur.*)

Figure 7.16. Improving a picture of Comet Hyakutake by digital image processing. The original slide (top) was scanned, a dark spot was retouched out, unsharp masking was performed to bring out detail, and contrast was adjusted.

You can also adjust the contrast of the picture to set the black and white levels where you want them and to control the gradation in between.

Once you have created a digital image, not only can you print it out – often making a far better print than could be made in a darkroom – but you can also share it with others on the World Wide Web. Some of my pictures are on display at http://www.covingtoninnovations.com; I also maintain web links to many other astrophotographers' work.

# Chapter 8
# Troubleshooting

This chapter gives quick solutions to a number of common problems that are likely to puzzle a new telescope owner. This list is far from complete. For more information, see your instruction book, as well as Chapters 10–12 and the websites mentioned there.

## 8.1 Electrical and computer problems

### Telescope is electrically dead

Check all fuses. On the LX200, there is a fuse inside the connector panel as well as in the power cable.

Check keypad and declination cables too. If deprived of any of its necessary connections, the computer will not initialize normally and the telescope may seem to be electrically dead.

Check for a coaxial power connector that is the wrong size. The diameter of the central pin is smaller on Celestron than on Meade telescopes. The Meade connector fits Celestron telescopes but gives a loose connection.

### Computer hangs or resets at random moments

This is often caused by a loose connection somewhere in the power system; inadequate battery voltage; or electrical noise on the power line, such as poor filtering of an AC-operated power supply.

If using internal batteries, check the battery contacts and try using an external power supply. Rechargeable lead-acid batteries are the gold standard for power supplies; they produce very clean, well regulated DC.

Make sure coaxial DC connectors are the right type and are firmly plugged in. On the NexStar 5 and some others, the central pin of the socket consists of two tines that you can spread apart with a screwdriver to get a better connection.

Loose connections in the keypad cable are also a possibility, as are accidental short circuits inside the telescope.

### LX200 fails to retain site data and other settings
Except for the date and time, these settings are stored in a flash memory chip that does not require battery backup. I have encountered a couple of cases of this chip being defective. It is a 24C04A or equivalent (check your telescope to be sure) and is in a socket on the main circuit board.

### LX200 fails to retain date and time
The clock-calendar of the LX200 is powered by a lithium battery just inside the connector panel. This battery will require replacement after a few years.

### NexStar fails to retain site data
On some NexStar 5 and 8 telescopes, only site numbers 0, 4, and 8 are usable. Owing to a firmware bug, attempting to use other site numbers will overwrite the memory locations in between.

## 8.2   Keypad problems

### Keypad display is blank (no light)
On LX200 and NexStar telescopes, the display illumination can be switched off. On the Autostar, it switches off automatically after several minutes with no keys being pressed. To restore it, on the LX200 or Autostar press MODE; on the NexStar, see p. 191.

   If the display is completely dead, the connection to the keypad may be loose. Turn the power off and check both ends of the keypad cable.

### Keypad display is very dim
Keypad brightness is adjustable. Check the instructions for your specific telescope.

   The Autostar display normally becomes very dim after you enter the time of day. That is because it is switching to the brightness level you have selected for nighttime use. You can select a brighter level.

### LX200 keypad display is blank but illuminated
The telescope is displaying the distance to the object as a bargraph (p. 166), and the distance happens to be practically zero. Press MODE and/or GO TO to return to a familiar display.

### ENTER key is not recognized
On Meade telescopes, a "long Enter" (over $\frac{1}{2}$ second on the LX200, over 2 seconds on the Autostar) is different from a regular Enter keystroke. If a regular Enter is what you need, a long Enter is not acceptable. Try pressing the key for a shorter time.

### Autostar does not recognize keystrokes

The Autostar scans its keyboard rather slowly, and you must type slowly, pressing each key for about $\frac{1}{4}$ second or more. Keys are recognized when you release them, not when you press them.

### NexStar does not recognize keystrokes

Most keys on the NexStar can be used only when the NexStar Ready prompt is being displayed. Press UNDO to get to that prompt.

### Menu selections do not seem to work

On the Autostar, a menu item is not actually selected unless > appears in front of it. Press ▲ or ▼ at the bottom of the keypad to make the selection, then press ENTER. Just getting the item onto the screen is not enough; if you scroll to an item and press MODE, the item is not selected.

### Incorrect information on keypad display

Concerning errors in the LX200 firmware, see p. 162.

The Autostar display uses abbreviations that are sometimes cryptic, such as THE TAU for Theta ($\theta$) Tauri.

When used on an equatorial wedge, the NexStar 5 reckons altitude and azimuth from the plane of its own base rather than the horizon. When asked for the altitude of an object, it displays something close to the declination. More generally, when used on any nonlevel tripod, the NexStar 5 reports altitude incorrectly.

### Text scrolls horizontally too fast to read

On the Autostar, the scrolling speed can be adjusted with the ▲ and ▼ keys at the bottom of the keypad.

### Keypad beeps or squeals

Many telescopes have a built-in timer and/or alarm clock with a beeper in the keypad.

### Scrolling down through a long menu is tiresome – there has to be a better way!

There is. The menus generally wrap around. To go straight from the first item to the last item, scroll *up* one step.

Similarly, since most star names begin with A (from Arabic *al-*), the quickest path to Polaris or Regulus in a list of named stars is to scroll up, wrapping around to the bottom of the list and then moving upward.

## **8.3**  Motor and slewing problems

### Motors whir but telescope does not move

If you hear the motors running when you press the slewing buttons, but the telescope does not move, its brakes (clutches) are probably unlocked. The motors control the telescope only when the brakes are locked.

The telescope may also be at a limit of its motion. For example, the Meade ETX-90 will not revolve around and around in azimuth; it reaches a limit.

### Starting the alignment process, telescope points far away from intended star

If the telescope says it is slewing to a star and immediately points at the ground, or at a part of the sky far away from that star, it probably did not start in the correct starting position ("home position").

The telescope may be set for equatorial (polar) mode while on an altazimuth mount or vice versa.

In equatorial mode, Meade ETX telescopes can be put on their wedge or tripod backward, leading to confusion. (The connector panel should face west.) Newtonians can be placed in the cradle backward.

See also the next item.

### Large errors in pointing

The telescope may have been aligned on the wrong star; repeat alignment procedure.

The date and time may be wrong; check clock/calendar and backup battery inside connector panel. Make sure the date is entered in the correct format (month/day/year unless otherwise specified).

Time zone and other site data may be wrong. Note that U.S. time zones and longitudes are negative on most telescopes, but positive on the LX200. The LX200 north/south switch (for northern or southern hemisphere) may be in the wrong position.

On Autostars and NexStars, make sure the right model of telescope is selected on the keypad. The correct setting should be retained permanently in the computer's memory, but it is worth checking when problems are observed.

The telescope may be slipping on the right ascension or declination axes; check the brakes (clutches). Once the telescope is aligned, it can only be moved electrically; do not unlock the brakes. (The Celestron Ultima 2000 is an exception.)

The telescope may have slipped because of trying to slew to an impossible position. This happens, for example, when the larger NexStars try to aim below the horizon; they are designed to slip on the altitude axis rather than break the mechanism. Turn the telescope off and realign.

The coordinates of a few objects are listed incorrectly in the built-in firmware of some telescopes. For example, when told to slew to M10, some NexStars go to a point in Ara about $40°$ south of the correct position. Try going to the object by right ascension and declination rather than by name.

For unknown reasons, the NexStar 5 occasionally goes to a wildly incorrect position when told to slew to an object, or moves in the right direction but stops halfway. Selecting the same object again sends it to the correct position.

### Small errors in pointing

Factors that affect pointing accuracy are discussed on p. 32. Bear in mind that extreme accuracy is not achieved with portable equipment. There are too many things that can flex.

Pointing will always be most accurate near the stars on which the telescope was aligned. To improve pointing accuracy in a particular area of the sky, sync on a star there.

Do not judge a telescope by how well it finds the alignment stars *before* you have performed the alignment. At that point, it is working blind and is completely at the mercy of the leveling of the tripod and the initial position of the telescope. After you sight on two stars, the telescope knows the position of the whole sky.

Autostar telescopes require a "Train Drive" procedure when first used and every few months thereafter. An inaccurate Autostar may also require the "Calibrate Motors" procedure.

In equatorial (polar) mode, the LX200 requires extremely precise polar alignment in order to find objects accurately. A quick workaround is to sync on a star in the part of the sky that is of interest before going to a difficult object.

See also the previous item (large errors in pointing); you may have a mild case of one of the problems described there.

### Telescope goes to wrong object

On the Autostar, you must select an object by pressing ENTER before pressing GO TO to slew to it. If you bring up an object on the screen and press GO TO, the telescope will go to the previous object, which is still selected.

### Excessive noise

When slewing at high speed, the LX200 sounds somewhat like a coffee grinder. Noise can be reduced by selecting a slower slew rate (see p. 165). Excessive noise in an older LX200 may indicate that a worm gear needs greasing and/or aligning.

The ETX-90 normally makes groaning or creaking noises while tracking (i.e., while "standing still", remaining pointed at an object as the Earth rotates). Most other telescopes are almost silent while tracking.

### Mechanical looseness

On lower-priced telescopes, some looseness in the mount is normal, especially when power is turned off. The pointing system compensates for the looseness through its backlash compensation settings and in other ways.

### Poor tracking

If the telescope does not track an object after finding it, the problem is probably that tracking is turned off ("land mode") or the wrong tracking mode is selected (altazimuth or equatorial, and if equatorial, northern or southern hemisphere).

On Autostars and NexStars, make sure the right model of telescope is selected on the keypad. The correct setting should be permanent, but it is worth checking when problems are observed.

On low-end telescopes, tracking is unfortunately sometimes poor. Most telescopes track much more smoothly in equatorial mode than in altazimuth mode.

Some NexStar 4's have a known firmware bug causing very poor tracking. A fix is available from Celestron.

Very irregular tracking on the LX200 can be caused by random or incorrect data in the Smart Drive (periodic-error correction) memory. Clear the Smart Drive and, if you wish, retrain it. If the problem recurs, try replacing the 24C04A chip mentioned on p. 123.

### Telescope does not move when tracking sky objects

It is normal for a telescope to look as if it isn't moving when it is tracking the sky (compensating for the Earth's rotation). The only way to tell whether the telescope is tracking the sky is to look in the eyepiece.

### Telescope refuses to slew to certain parts of the sky

This is a common problem with Meade ETX telescopes and is caused by starting with the telescope in the wrong position. It is not enough to point the telescope north; it must also be within 180° of its anticlockwise limit. That is, you should turn it anticlockwise as far as it will go; then turn it clockwise to north; then turn it on.

Other telescopes may have similar problems. Check clock, calendar, and site information and realign.

On the LX200, another possibility is a loose connection in the declination cable.

### Telescope goes "the long way round," rotating nearly 360° to make a short movement

This is a protective measure to keep the telescope from tangling its cords (including the internal cords in the Meade ETX). On NexStar telescopes, cord wrap protection can be turned off.

If, with this in mind, the actions of the telescope still do not make sense, check whether the telescope was initialized in the wrong starting position (pointing south vs. pointing north) or is set for the wrong hemisphere (northern vs. southern).

### After finding the object, telescope moves off it

On the Autostar, pressing GO TO after arriving at the object initiates a square-spiral search of the sky around it. Press MODE to stop.

### Incorrect motion when you press an arrow button

If motion begins with a sudden jerk, the backlash compensation is set too high. See the instructions for your specific telescope and set a lower value.

If there is a delay before motion begins, or even a brief movement in the wrong direction, then the backlash compensation is set too low. Unless the problem is severe, this is not a bad thing; it is always better to have too little compensation than too much.

Note that on the NexStar 5 and some others, the up and down arrows are reversed at slew rates 1 to 6, but not at rates 7 to 9. The idea is that you will use the high rates when looking through the finder and the low rates when looking through the eyepiece (with diagonal). In each case the arrows should match what you see.

### Slewing arrow buttons have no effect after connecting autoguider

The LX200 resets itself to the slowest slewing speed ("Guide") when you connect a CCD autoguider or when there is any activity on the CCD autoguider port. However, the lights on the keypad still indicate that a higher speed is selected. You can select a different speed manually by pressing the appropriate button.

### "Runaway" (motion that does not stop)

In general, this indicates a failure of the encoders (p. 21) so that the motors can still move but cannot report their position.

Sometimes, however, runaway is caused by a loose connection. On the LX200, check the declination cable. Check also for electrical problems, such as inadequate battery voltage, loose connections, or even a metal cover plate accidentally touching connections inside.

Some cases of runaway on the LX200 have been traced to overheating of voltage regulator ICs when operating from an 18-volt supply in warm weather. The ICs shut down for safety, leaving the encoders inoperative. Try a 12- or 15-volt supply.

See also "Computer hangs or resets at random moments" (p. 122), since the same problems can also cause runaway.

### Unduly difficult polar alignment

If you cannot get consistent, accurate polar alignment of an equatorial mount even by using iteration or the drift method (p. 48), the mount is probably shifting while you are unaware of it. Look for loose bolts anywhere in the mount or tripod.

With many wedges, a large shift can occur while you are tightening the lock-down bolts; to guard against this, tighten them carefully while looking through the eyepiece. Also, make sure the lock-down bolts are quite loose before adjusting the wedge, so that there will be no pent-up tension when you tighten them again.

### Cannot lock R.A. brake

On the LX200, the ETX, and others, the right ascension brake handle may not be able to move far enough to lock the brake. You can loosen the setscrew and reposition the handle on the shaft.

## 8.4 Optical problems

### No image in eyepiece (whole field is dark)

Some Meade ETX and Celestron NexStar models have a built-in flip mirror to control whether the image is formed at the eyepiece or at the camera port in the back.

### Image quality is poor

Understand that high-power eyepieces never give a crisp image; you are working near limits imposed by the laws of physics. For daytime testing, try a 25-mm eyepiece.

If the image seems to be constantly moving and boiling, the problem is unsteady air. Allow the telescope to come to thermal equilibrium with its surroundings after coming from indoors. Allow one hour per 18 °F (10 °C) of temperature change. Even after taking all these precautions, the air may still be too unsteady for high-power observing, especially right after the passage of a cold front.

Check collimation (see p. 70). This adjustment is meant to be performed by the user and should be checked whenever the telescope is used to view stars in steady air; readjustment is required every few weeks or months. The symptom of poor collimation is that stars look like comets or cones rather than round disks.

Do not test the telescope by viewing through a window. Window glass is not of high enough optical quality. Even an open window will cause air currents that distort the image.

### Dark spot in middle of image

A dark spot in the middle of the image, shifting as you move your head slightly, usually means you are using a low-power eyepiece in the daytime, when the pupil of your eye cannot take in the whole beam of light coming out of the telescope (see p. 88).

Normally, the dark spot is caused by the shadow of the secondary mirror. For a different reason, some eyepieces (especially early Naglers) produce a bean-shaped shadow that darts around the periphery of the image. Like the secondary-mirror shadow, this "kidney bean effect" is evident only when the pupil of the eye is constricted in bright light.

Occasionally, the only problem is that your eye is the wrong distance from the eyepiece, either too far or too close.

### Image shifts sideways while focusing

Some lateral image shift is inevitable in the Schmidt–Cassegrain and Maksutov–Cassegrain focusing mechanism. Image shift can often be greatly reduced by running the focuser all the way from one end of its range to the other several times to redistribute lubricants.

### Image will not hold focus

At high power, especially when lubricants are cold and stiff, the focus may continue to shift for a moment after you let go of the focusing knob. On the LX200 and similar telescopes, it is best to do your final focusing by turning the focusing knob *anticlockwise*. That way, you are tightening a spring rather than loosening it, and there will be no further settling after you let go.

### Diagonal prism is off center

This is a common problem with a batch of Celestron diagonals made around 1998, and it makes collimation impossible as long as the diagonal is in the system. It is easily fixed by taking the diagonal apart with a screwdriver, putting the prism firmly into its mount, and reassembling.

### Complete inability to focus with camera adapter; light, but no image

Not all camera adapters work with all telescopes. In particular, a camera body without a lens often will not reach focus with a Newtonian reflector, whose focal plane is deep within the eyepiece tube; use afocal coupling or eyepiece projection instead.

Also, if the camera has a lens, the telescope must have an eyepiece; you cannot put a digital camera (with nonremovable lens) into an adapter designed for a camera body without a lens.

### Focusing with camera adapter is difficult

At high magnifications and high $f$-ratios, the image does not snap into focus the way it would with an $f/2.8$ camera lens. Considerable skill is needed. Focusing an SLR camera is easiest if you use a plain matte focusing screen and focus on the Moon or a bright star. See p. 115.

**Part II**
Three classic telescopes

# Chapter 9
## Three that led the revolution

Around 1999 and 2000, when amateur astronomers adopted computerized telescope technology *en masse*, three telescopes led the revolution. They were the Meade LX200 (on the market since 1992), the Meade ETX-90 EC Autostar (introduced in 1999), and the Celestron NexStar 5 (2000).

The following chapters describe these telescopes in some detail. By now, none of them is still the manufacturer's latest and greatest. Technology is progressing so fast that new models appear almost every month.

But the older telescopes will still work as well as they ever did, and tens of thousands of them will remain in use for many years. The information in the next three chapters will help those who still use classic computerized telescopes, or are thinking of buying them, or simply want to know what they are like.

# Chapter 10
# Meade LX200

## 10.1 Introduction

This chapter describes the original Meade LX200 (Figure 10.1), which was made from 1992 to 2001. The newer LX200 GPS is optically and mechanically similar but has an enhanced version of the Autostar computer described in Chapter 12.

The information here is based on my experiences with an 8-inch LX200 purchased in 2000. I assume that you also have the Meade manual available for reference. This is not a complete guide to all the LX200's features.

This chapter is more detailed than the next two, for several reasons. The original LX200 was on the market for nine years, so there was plenty of time for the amateur community to learn all about it. All LX200s use similar firmware, and the total number of LX200s in use is very large, so this detailed information is useful to a large number of people. Finally, the original LX200 is at the end of its product life cycle, so there will be no further changes.

Even after it is discontinued, the LX200 will remain in wide use for many years. It is to computerized telescopes what the Nikon F is to cameras: an army of loyal users complain about its quirks but continue to trust it for serious work.

### 10.1.1 Evaluation of the LX200

Two useful features are conspicuously missing from the LX200: the ability to download software updates, and the ability to do a two-star alignment in equatorial mode to keep pointing accuracy from depending on polar alignment.

Apart from those minor drawbacks, which are corrected in the LX200 GPS, the original model LX200 is a tried and true performer. Its keypad is designed for experienced observers, and common operations, such as finding a star or planet, require considerably fewer keystrokes than on contemporary Autostar models (but see p. 193).

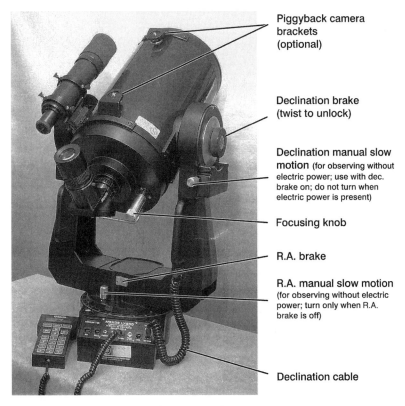

Piggyback camera
brackets
(optional)

Declination brake
(twist to unlock)

Declination manual slow
motion (for observing without
electric power; use with dec.
brake on; do not turn when
electric power is present)

Focusing knob

R.A. brake

R.A. manual slow motion
(for observing without electric
power; turn only when R.A.
brake is off)

Declination cable

Figure 10.1. Important parts of an LX200 telescope.

### 10.1.2 Related products

The 8- and 10-inch versions of the LX200 use the same firmware, and the instructions here are 100% applicable to them. The 7-inch Maksutov–Cassegrain and the 12-inch Schmidt–Cassegrain have slightly different firmware, since their tubes do not swing through the fork arms and must be blocked from slewing to impossible positions, but apart from that, there are no significant differences. The 16-inch Schmidt–Cassegrain has another version of the firmware with additional features.

Almost identical computers, with the same hand box, are used on some Meade German equatorial mounts supporting both refractors and reflectors.

### 10.1.3 Firmware versions

This chapter is based on my experiences with firmware version 3.34L, the current version since at least 1999. The firmware version is displayed on the keypad as the telescope is powered on. Do not be misled by a label on the backside of the keypad such as "Revision 3.21" describing the hardware version of the keypad itself.

Earlier versions of the firmware have a few inaccurate star positions, leading to less accurate alignment and pointing. The firmware can only be upgraded by obtaining replacement chips and/or circuit boards from Meade. Versions before 3.30 (1995) have substantially fewer features and should be upgraded.

### 10.1.4 LX200 websites

The definitive guide to the LX200 is the website of the Meade Advanced Products Users Group (MAPUG) at http://www.mapug.com, with links to other sites. LX200 telescopes are discussed on the MAPUG mailing list and the newsgroup alt.telescopes.meade.lx200. Meade's website, including online manuals, is http://www.meade.com.

## 10.2 Electrical requirements

*Make sure the telescope is switched **off** before connecting or disconnecting any cables, including power, keypad, and declination motor.*

To operate the telescope, you will need to plug in the declination motor cable, the keypad, and DC power.

The LX200 requires a power supply that delivers between 12 and 18 V DC and can supply 1.5 amperes continuously, 2.0 amperes momentarily. Actual current drain is about 0.8 A, with occasional bursts of higher current. Slewing requires only slightly more current than tracking.

Early model LX200s were specified to run on 12 volts, not 18. Do not use an 18-volt supply with an older LX200 unless you are sure it is permissible.

The power connector is a coaxial plug, 5.5 mm o.d., 2.5 mm i.d., center positive. When in doubt, measure the connector to ensure a good fit. In particular, $5.0 \times 2.5$-mm plugs will plug into the telescope but will not give a good connection.

*Caution:* SBIG CCD cameras use the same connector, but with the opposite polarity. Various other devices use similar connectors with various voltages and polarities. Do not mix them up. Applying power to the LX200 with the wrong polarity will cause damage.

The LX200 warranty requires that the power cable contain a 1.5-amp slow-blow fuse outside the telescope. An additional fuse is located inside the connector panel.

An 18-volt supply is recommended for faster slewing and more reliable tracking, particularly when the telescope is out of balance carrying photographic equipment. For ordinary operation of an 8-inch (20-cm) telescope, a 12-volt supply is quite adequate. Those using an 18-volt supply may prefer to set a maximum slew rate of 4 (on the TELESCOPE menu, p. 165) rather than the default of 8, to reduce noise and wear.

A useful accessory is the Meade 1812 voltage converter. This is a switching regulator that converts 12 volts to 18 volts. The 1812 can overheat if the input

Figure 10.2. The LX200 connector panel. Bargraph display at upper left is ammeter, showing current consumed by telescope.

voltage goes substantially below 12 V, because in order to continue delivering 18 V at the output, it must draw excessive current.

At the upper left corner of the connector panel (Figure 10.2) is a bargraph display of the current drawn by the telescope. Contrary to the label, the bars do not actually correspond to 100-mA units. On my telescope, each bar represents about 120 mA at 12 volts or 150 mA at 18 volts.

## 10.3 Keypad

The most important keypad functions are shown in Figure 10.3.

MODE is the "undo" or "back up" key. Press it whenever you want to back out of a menu without making a choice, or when you get lost in the menu system.

PREV (▲) and NEXT (▼) are used for making menu choices; N, S, E, and W are for slewing the telescope. W and E also move the cursor left and right when you are typing numbers in.

It is normal for the keypad to get warm during operation, particularly when using an 18-volt power supply. The heat comes from a voltage regulator that steps the supply voltage down to 5 V for the keypad. One backhanded advantage of the excessive heat is that it keeps the LCD display from fading out in very cold weather.

The keypad illumination is always uneven. A few LEDs inside its housing provide illumination for all the keys. Their brightness can be adjusted (see p. 164).

Note that the computer itself is not in the keypad – the keypad is only a terminal. The main computer is under the base of the telescope.

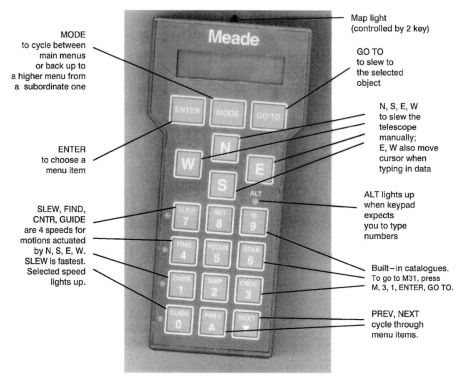

Figure 10.3. The LX200 keypad.

### 10.3.1 Direction of movement

The N, S, E, and W buttons are laid out to match the view through the eyepiece with a diagonal. For example, to move to an object that is to the lower right of the center of the field, push the buttons at the bottom and the right (S and E respectively).

The TELESCOPE menu gives you the ability to reverse N and S and to reverse E and W.

### 10.3.2 How to enter negative numbers

When entering site latitudes, time zone numbers, and declinations, you will sometimes need to type a negative number.

To do this, use the W key to move the cursor leftward onto the + or − sign that you wish to change. Then press PREV for + or NEXT for −. Finally, use E to move the cursor back into the numeric field, and type the digits.

### 10.4 Operation without electricity

The LX200 is one of the few computerized telescopes that are fully usable without electric power (except of course that the drive motors do not run). Operation is very similar to the original Celestron 8 or any other noncomputerized telescope on a similar mount; the main difference is that the LX200 is not as well balanced, but that can be corrected with counterweights.

There are right ascension and declination slow motion knobs and setting circles. The right ascension slow motion works with the brake off; the declination slow motion works with the brake on. The right ascension setting circle can be rotated manually to match the sky.

## 10.5 Motorized operation without alignment

In "land mode", you can slew (aim) the telescope electrically (at four speeds) without requiring any site data or alignment. This is a quick way to get acquainted with the telescope. It is also the usual way to use the telescope to view land objects (mountains, ships, etc.).

**It is strongly recommended that you try out land mode before trying anything more advanced.**

To get into land mode:

(1)     Turn power on. After the computer initializes, you will see this menu:

```
→TELESCOPE
  OBJECT LIBRARY
```

If object library is shown in lowercase letters, the telescope is already in land mode and you can stop. Otherwise, proceed . . .

(2)     Select TELESCOPE. That is: use PREV and NEXT to put the little arrow next to TELESCOPE (if it isn't there already), then press ENTER.

(3)     You will now see this menu:

```
→1)SITE
  2)ALIGN
```

Press NEXT to put the arrow next to ALIGN, then press ENTER.

(4)     The next menu begins with

```
→ALTAZ
  POLAR
```

but you can scroll down with NEXT to see a third option, LAND. Do so. When the menu looks like this:

```
  POLAR
→LAND
```

press ENTER.

The telescope is now in land mode. It will remember this setting even when turned off, until you explicitly choose a different mode.

In land mode, you can move the telescope up, down, left, and right with the N, S, W, and E buttons.

## 10.6 Controlling the slewing speed

You can choose from four slewing speeds by pressing SLEW (fastest), FIND, CNTR, or GUIDE. The last of these is very slow and is mainly for photography. The currently selected speed is indicated by an LED next to the button that selects it.

The fastest speed can be further selected by a setting on the TELESCOPE menu. The default is 8, but many people prefer 4, which is considerably quieter. This is also the speed at which the telescope moves when told to go to a celestial object.

## 10.7 Entering date, time, and site information

You need only enter this information once; the clock inside the telescope runs accurately even when powered off, though it should be checked every few weeks.

There is room in nonvolatile memory for up to four site locations. You can give each site a name. "Unknown site" operation is also supported.

### 10.7.1 Setting the time

You can use either your local time or Universal Time (UT, also called Greenwich Mean Time, GMT). It is a good idea to check the clock's accuracy every few weeks.

To set the time, do the following:

(1)   Press MODE until the display looks like this:

```
LOCAL = 11:22:33
 SIDE = 22:33:44
```

Here SIDE denotes sidereal time, which is calculated for you; you will only be entering the local time.

(2)   Press and hold ENTER until you hear a beep. (This is a keystroke called **long ENTER**.)

(3)   Type in your local time using the digits on the keypad. *Use 24-hour format.* To make corrections, move the cursor with E and W. When finished, press ENTER.

(4)   You will now see a display like this:

```
Hours from GMT:
+08
```

This is the difference between your time zone and GMT, *but the sign is the opposite of the usual one.* On the LX200, Eastern Standard Time is +05, even though it is officially Time Zone −5, and Central European Time goes into the telescope as −01 even though it is officially +1.

Enter the correct value. If you are in Europe, Africa, Asia, or Australia, you will need to change the + to −; do this by moving the cursor to it and pressing NEXT for negative or PREV for positive.

If you are using GMT, you should enter +00, of course, and make sure the date as well as the time is correct for GMT.

### 10.7.2 Setting the date

If you chose to set the clock to Greenwich Mean Time, be sure also to use the GMT date (for example, 8 p.m. Eastern Standard Time is 01:00 GMT the following day).

To set the date, do the following:

(1)    Press MODE one or more times until you see the clock display:

```
LOCAL = 11:22:33
  SIDE = 22:33:44
```

(2)    Press ENTER. The display switches to the date:

```
DATE = 01/02/03
```

Press and hold ENTER until you hear a beep; this will enable you to type in a new date. Do so and press ENTER.

(3)    The telescope will briefly recalculate planet positions, displaying the message:

```
Updating
planetary data ...
```

while it does so.

(4)    When the recalculation finishes, you can press MODE to get back to other menus.

### 10.7.3 Entering site latitude and longitude

#### Getting your information into the LX200's format

Latitudes are positive if north, negative if south. *Longitudes are reckoned differently by the LX200 than by standard reference books.* The LX200 does not distinguish east and west longitudes. Instead, it treats longitude as running westward from Greenwich (0°) all the way up to 360°.

West longitudes (as in America) go into the telescope as positive numbers; thus Los Angeles, at 119°30′ W, goes in as 119°30′.

East longitudes are subtracted from 360°. Thus Cambridge, England, at 0°08′ E, goes in as 359°52′.

### Setting the hemisphere switch

Set the N/S switch on the connector panel (Figure 10.2, upper right) to N if you are in the northern hemisphere and S if you are in the southern hemisphere. You may want to tape it in place so that you will not accidentally flip it when turning power on or off.

### Storing and naming an observing site

The computer has enough memory to store the latitudes and longitudes of four sites, each of which you can label with a three-letter name. This information is stored permanently until you change it.

All sites are assumed to be in the same time zone. If you travel a lot, you may choose to run the clock on GMT as described above, or pick one time zone and use it for astronomical purposes everywhere you go.

To store latitude and longitude information for a site, do the following:

(1)   Instead of land mode, put the telescope into POLAR or ALTAZ mode. See the previous section for instructions. You do not have to perform an alignment; just select a mode other than LAND.

(2)   Press MODE until you see the usual opening menu:

```
→TELESCOPE
 OBJECT LIBRARY
```

Choose TELESCOPE and press ENTER.

(3)   On the menu

```
→1)SITE
 2)ALIGN
```

choose SITE and press ENTER.

(4)   You will now see the menu

```
→1)AAA  ✓
 2)AAA
 3)AAA
 4)AAA
 5)UNKNOWN
```

not all of which is visible at once, of course; scroll up and down with PREV and NEXT.

The check-mark (✓) indicates which site is currently selected. Even when powered off, the telescope will remember which site you are using until you change it.

(5)   Choose the site you want to edit (probably the first AAA). Initially, all four site memories are named AAA and contain the latitude and longitude of Meade's factory in Irvine, California (about 117° west, 33° north).

(6)   Press ENTER. That selects the site, putting the check-mark next to it.

(7)   Press and hold ENTER until you hear a beep. You will get the flashing cursor that indicates that editing is possible.

(8)   *Give the site a memorable three-letter name.* Move the cursor from letter to letter with E and W. Change each letter by pressing PREV or NEXT to cycle through the alphabet. When you have entered the name, press ENTER.

(9)   Now the display looks something like this:

```
LAT =+33°35'
LONG =117°42'
```

Type in your latitude.

If you are in the southern hemisphere, you will need to change the + to −; do this by moving the cursor to it and pressing NEXT for negative or PREV for positive.

When you have the correct latitude entered, press ENTER.

(10)   Type in your longitude the same way, remembering that western longitudes are positive, and eastern longitudes are subtracted from 360° as described earlier.

(11)   Press ENTER. You're done!

The telescope will retain the site information permanently in flash EEPROM. You can store up to four sites and select them from the menu.

## 10.8 Aligning the telescope on the sky

### 10.8.1 Altazimuth mode

#### Zero-star *approximate* alignment (known site)

Assuming correct site and time information have been entered, and the correct site has been selected from those stored in the telescope, you can do a *very rough* alignment of the telescope without sighting on any stars, as follows:

(1)   Place the telescope on the tripod with the connector panel toward the north.

(2)   Level the tripod carefully (*critical*). The built-in bubble level is not especially accurate; a carpenter's level on top of the tripod is better.

(3)   Unlock the right ascension brake, aim the telescope due south (*critical*), then lock the right ascension brake.

(4)   Unlock the declination brake, move the tube so that the declination setting

circle (Figure 4.12, p. 54) reads 0°, and relock the brake.

(5)    Turn power on.

(6)    Make sure the correct site is selected. From

```
→TELESCOPE
  OBJECT LIBRARY
```

choose TELESCOPE, then SITE, then select the site and press ENTER.

The telescope is now *approximately* aligned on the stars, but doesn't know it. Thus, you cannot type in a right ascension and declination and go to them. But you *can* select a library object (star or deep-sky object) and go to it. After you have done this at least once, you will also be able to type in right ascensions and declinations.

   *Don't expect much accuracy;* your estimate of due south was probably off by several degrees. However, you can make the alignment much more accurate by syncing on any celestial object (see p. 154).

### One-star alignment (known site)

Follow the procedure just given, except that aiming the telescope due south is not as critical. (Leveling the tripod remains very important.) Then continue as follows:

(1)    Press MODE as many times as needed to get to the initial menu:

```
→TELESCOPE
  OBJECT LIBRARY
```

Select TELESCOPE and then ALIGN:

```
  1)SITE
→2)ALIGN
```
(Press ENTER.)

Then choose ALTAZ:

```
→1)ALTAZ  ✓
  2)POLAR
```

If ALTAZ is already checked (✓), press ENTER just once. Otherwise, press ENTER to put the check-mark next to ALTAZ, then again to start the alignment procedure.

(2)    At the menu

```
1 Star or
2 Star Alignment
```

press 1 on the keypad.

(3)    The telescope will step you through the alignment procedure, as follows:

```
Level base, then
press ENTER
```
(Check the leveling one last time.)

```
Press ENTER,then
pick align star
```
(Press ENTER.)

```
→ACHERNAR
  ACRUX A
```

(4)    Using PREV and NEXT, scroll down to a star you can see that is in the sky, but not directly overhead. Select it by pressing ENTER. You'll get a message such as:

```
Center DENEB
then press ENTER
```

Aim the telescope at the star and center the star in the eyepiece. You can do this by slewing with N, S, E, and W at any of the four speeds, and/or by loosening the brakes and retightening them after moving the tube.

The telescope is now aligned on the celestial sphere and will track it automatically with its motor drive. From now on, make all movements electrically; leave the brakes tight.

### Two-star alignment (known site) (RECOMMENDED METHOD)

Proceed as above, but select two-star alignment. You will need to center two stars (the second one must be centered electrically, not by loosening the brakes). However, leveling of the tripod will be much less critical.

This is the mode that normally gives the best performance. The celestial sphere is located by the two stars, not by the levelness of the tripod.

Also, it is more foolproof. If you misidentify one or both stars, the telescope will usually complain because they will be the wrong distance apart.

For advice about which stars to use, see p. 28. You must find the first star manually; if you press GO TO after selecting the second star, the telescope will slew to its approximate location.

The telescope still assumes the tripod is level in order to determine where the horizon is, so that it can correct for the bending of starlight by atmospheric refraction near the horizon.

### Two-star alignment (unknown site)

Two-star alignment works equally well when site information is not available; in that case you simply choose UNKNOWN from the menu of sites, then proceed as just described.

In this case there is no correction for atmospheric refraction because the telescope does not know exactly where the horizon is.

## 10.8.2 Equatorial mode

**Zero-star *approximate* alignment (polar-aligned mount)**
If the telescope is on a mount that is already accurately polar-aligned (in practice, 0.5° accuracy is considered good), and accurate date, time, and site information have been stored in the telescope, you can align it roughly with the stars without actually sighting any objects, as follows:

(1)   Unlock the declination brake, move the tube so that the setting circle reads 0°, and re-lock the brake.

(2)   Unlock the right ascension brake, move the tube so that the telescope is pointing due south (on the meridian, zero hour angle, Figure 10.5), and re-lock the brake. The starting position is shown in Figure 10.4.

(3)   Turn power on.

*Don't expect great accuracy* until you sync on an object.

**One-star equatorial alignment (polar-aligned mount)**
As above, except that your next move is to GO TO any library object (a star, cluster, nebula, or galaxy), then slew the telescope with N, S, E, and W until the object is actually centered in the eyepiece, then press and hold ENTER until you hear a beep.

This is called "syncing on" the object. For best results, choose an object that is moderately high in the southeast or southwest sky (not directly overhead).

*This is the recommended technique when using a permanent mount that is accurately polar-aligned.* Note that the ALIGN menus are not involved at all!

**Alignment on Polaris and one additional star (RECOMMENDED)**
Ordinarily, you will *not* be sure that your mount is accurately polar-aligned before putting the telescope on it. Instead, proceed as follows:

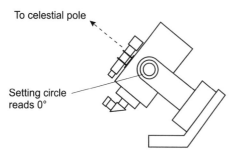

Figure 10.4. Starting position for LX200 on a mount that is already polar-aligned.

Figure 10.5. Hour angle zero means that the two outer marks are lined up. The position of the setting circle between them is irrelevant.

(1)    Polar-align your mount roughly (see pp. 43–47). One way to do this is to set your wedge to the latitude of your site, level the tripod, and place it so that the wedge is directly south of center.

(2)    Turn the power on. From

```
→TELESCOPE
  OBJECT LIBRARY
```

press ENTER and then choose ALIGN, then POLAR. If POLAR is not already checked (✓), you will need to press ENTER twice to start the alignment sequence.

(3)    At the prompt:

```
Move to Dec 90,
HA 0, Press ENTER
```

unlock the right ascension and declination brakes. Set the telescope to declination 90° (pointing directly away from its base) and hour angle zero (aiming directly away from its connector panel; line up the mark on the revolving part of the base with the mark on the stationary part, as in Figure 10.5). Re-lock the brakes. The correct starting position is shown in Figure 10.6.

(4)    Press ENTER. The telescope will slew to the computed position of Polaris, which is not directly on the pole. (This may be a very awkward and annoying

147

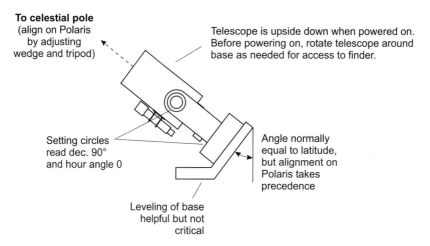

**To celestial pole**
(align on Polaris
by adjusting
wedge and tripod)

Telescope is upside down when powered on.
Before powering on, rotate telescope around
base as needed for access to finder.

Setting circles
read dec. 90°
and hour angle 0

Angle normally
equal to latitude,
but alignment on
Polaris takes
precedence

Leveling of base
helpful but not
critical

Figure 10.6. Starting position on a mount that is to be polar-aligned with help from the LX200's computer.

operation, since the telescope will aim north and then twist around its base; the finder may even end up almost inaccessible. Users of Telrad finders have been known to install a second finder base on the bottom of the telescope in case that happens.) Then you'll get this prompt:

```
Center POLARIS
then press ENTER
```

(5)     *By adjusting the mount, not the telescope,* get Polaris centered in the field. Depending on how far you're off and how your mount works, you may need to move the tripod slightly, lengthen or shorten tripod legs, and/or adjust altitude and azimuth on your wedge.

> *Hint:* At this point it no longer matters whether your tripod is perfectly level. Your goal is to get the telescope's polar axis pointing in the right direction. The bubble level was only a shortcut to help you get roughly there.

When Polaris is accurately centered, press ENTER.

(6)     The telescope will choose an alignment star (yes, it's smart enough to do that!) and slew to it. You'll get a prompt such as:

```
Center SPICA
then press ENTER
```

Center the star using N, S, E, and W, then press ENTER.

If the star that the telescope chooses is hidden by a tree, press ENTER anyway, then go to any known star and sync on it.

Congratulations! Your telescope is lined up on the stars.

**Refining polar alignment**

Methods for refining the polar alignment are described beginning on p. 48. Because pointing accuracy in equatorial mode depends on polar alignment, the use of iterative refinement or the drift method is strongly recommended.

## 10.9 Finding objects by coordinates

### 10.9.1 Slewing to a given R.A. and declination

Press MODE until R.A. and declination are displayed:

```
RA = 01:22:33
DEC = +23°45:00
```

Press GO TO.

Type in the desired right ascension and declination. Use E and W to move the cursor left and right. To change the sign to + or − as needed, move the cursor to it and press PREV or NEXT.

Finally, press ENTER and the telescope will slew to the object.

### 10.9.2 Slewing to a given altitude and azimuth

As above, but when the R.A. and declination are displayed on the keypad, press ENTER to switch to altitude and azimuth, and then proceed. Note that the LX200 reckons azimuth in an unusual way (p. 10).

### 10.9.3 Dealing with decimal minutes

Many catalogues give right ascension in hours, minutes, and tenths of minutes (e.g., 12:37.2), but the LX200 expects you to enter it as hours, minutes, and seconds (12:37:12).

The conversion is very simple: just multiply the tenths of minutes by 6, thus:

| | |
|---|---|
| $0.0^m = 00^s$ | $0.5^m = 30^s$ |
| $0.1^m = 06^s$ | $0.6^m = 36^s$ |
| $0.2^m = 12^s$ | $0.7^m = 42^s$ |
| $0.3^m = 18^s$ | $0.8^m = 48^s$ |
| $0.4^m = 24^s$ | $0.9^m = 54^s$ |

Recall that a minute of right ascension is 15 times as big as a minute of declination. Normally, seconds of declination can be ignored.

## 10.10 How to interrupt a slewing movement

If the telescope is on its way to an object and you need to stop it for some reason, perhaps to keep it from tangling cords or bumping into an obstacle, press GO TO while the telescope is moving. Alignment is not lost when you do this.

## 10.11 Finding deep-sky objects using the built-in catalogues

### 10.11.1 M (Messier) Catalogue

The easiest part of the Object Library to use is the Messier catalogue because all you have to do is press M. You get the prompt:

```
M object:
—
```

Type in the M number, hit ENTER, and the object is selected. You can then cycle through three displays of information about it, such as these, by pressing ENTER:

```
 M96     VG GAL
MAG 9.2 SZ   7.1'
```

(This means the object is M96, a 'very good' [bright] galaxy, mag. 9.2, size 7.1'.)

```
RA  = 10:46.7*
DEC = +11°49'
```

(Here * means these are coordinates of the object, not the telescope.)

(This is a bargraph of the distance in R.A. and declination from the current position of the telescope.)

To go to the selected object, press GO TO.

### 10.11.2 NGC, IC, and UGC

The LX200 also contains three more catalogues of deep-sky objects:

- The complete *New General Catalogue* (NGC) containing 7840 objects.
- The complete *Index Catalogue* (IC), a supplement to the NGC.
- The *Uppsala General Catalogue* (UGC) of 12 921 galaxies. (The "UGCA" objects from a later supplement are not included.)

These are all accessed with the CNGC key; you must choose which catalogue to use, and it remains selected until you choose another one. Here's the procedure:

(1)    Press CNGC. You'll get a prompt such as:

```
NGC object:
—
```

but the first line *may* say IC or UGC instead of NGC.

(2)    If the currently selected catalogue is the one you want to use, type in the number and hit ENTER, then proceed just as with the Messier catalogue.

(3)    If you want to choose a different catalogue, press ENTER without typing a number. You'll get the menu

```
→NGC  ✓
  IC
  UGC
```

(not all visible at once).

(4)    Scroll down to the catalogue you want to select and press ENTER to place the check-mark ( ✓ ) next to it. It will remain the default catalogue for the CNGC key until you select a different one.

(5)    Press MODE to get back to the prompt for the object number.

## 10.12 Finding stars using the built-in catalogues

### 10.12.1 Named stars

The LX200 can identify 34 stars by name. They are:

| | | |
|---|---|---|
| Achernar | Bogardus | Mira |
| Acrux | Canopus | Polaris |
| Albireo | Capella | Pollux |
| Alkaid | Castor | Procyon |
| Aldebaran | Deneb | Regulus |
| Alnilam | Denebola | Rigel |
| Alphard | Diphda | Sirius |
| Alphekka | Enif | Spica |
| Altair | Fomalhaut | Vega |
| Antares | Hadar | |
| Arcturus | Hamal | |
| Betelgeuse | Markab | |

To go to any of these stars:

(1)     Press STAR. You'll get a prompt:

```
STAR object:
─
```
(or possibly SAO object or GCVS object)

(2)     Press ENTER without typing anything.
        This will take you to the list of alignment stars:

```
→ACHERNAR
  ACRUX A
  ALBIREO
  ALKAID
  ⋮
```

(3)     Scroll down to the star you want, and press ENTER.
(4)     Press GO TO to actually slew to the star.


### 10.12.2 STAR, SAO, and GCVS numbers

The LX200 contains three star catalogues:

- The 351-item Meade STAR catalogue, inherited from earlier-model LX200s that only had room for a few hundred stars. It contains 250 relatively bright stars used mainly for alignment, plus 100 interesting doubles, plus Sigma Octantis (the south pole star, STAR 351).
- 15 928 stars from the *Smithsonian Astrophysical Observatory Star Catalog* (SAO for short), down to magnitude 7.0. This is not the complete SAO catalogue, but it's enough to be useful.
- 21 604 stars from the Moscow *General Catalogue of Variable Stars*, down to magnitude 16.6 (at minimum). This is not the complete catalogue, since some stars with extremely faint minima – mainly novae – are excluded; but it contains nearly every known variable star.

You must select the STAR, SAO, or GCVS catalogue in the same manner as for the NGC, IC, and UGC deep-sky catalogues. Specifically:

(1)     Press STAR. You'll get a prompt such as:

```
STAR object:
─
```

but the first line *may* say SAO or GCVS instead of STAR.

(2)    If the currently selected catalogue is the one you want to use, type in the star number and hit ENTER to select the object, then GO TO to slew to it.

(3)    If you want to choose a different catalogue, press ENTER without typing a number. You'll get the menu

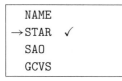

```
   NAME
 → STAR   ✓
   SAO
   GCVS
```

(4)    Scroll down to the catalogue you want to select and press ENTER to place the check-mark (✓) next to it. It will remain the default catalogue for the STAR key until you select a different one.

       (Note that you cannot actually select NAME. Instead, scrolling to NAME and pressing ENTER will take you directly to the list of names.)

(5)    Press MODE to get back to the prompt for the star number.

Instead of the traditional letter designations for variable stars (such as R Leonis), the LX200 uses GCVS designations in numerical form. The first two digits identify the constellation starting from Andromeda in alphabetical order. The last four digits stand for the position of the letter designation in the sequence R, S...Z, RR, RS...ZZ, AA, AB...QZ, and finally V334, V335, and so on, skipping the letter J wherever it would occur. Thus R Aquilae is GCVS 050001 (fifth constellation, first star), and V1024 Sagittarii is 721024. Tables 10.1 and 10.2 can be used to look up the numbers, or you can use the tables in the LX200 manual.

## 10.13 Finding the Moon and planets

The Moon and planets are accessed through the STAR catalogue, as follows:

        STAR 901 = Mercury
        STAR 902 = Venus
        STAR 903 = The Moon
        STAR 904 = Mars
        STAR 905 = Jupiter
        STAR 906 = Saturn
        STAR 907 = Uranus
        STAR 908 = Neptune
        STAR 909 = Pluto

Early versions of the firmware omit the Moon.

## Table 10.1 *GCVS constellation numbers*

The full GCVS number consists of the constellation code followed by the star designation (next table).

| | | |
|---|---|---|
| 01 Andromeda | 31 Cygnus | 61 Pavo |
| 02 Antlia | 32 Delphinus | 62 Pegasus |
| 03 Apus | 33 Dorado | 63 Perseus |
| 04 Aquarius | 34 Draco | 64 Phoenix |
| 05 Aquila | 35 Equuleus | 65 Pictor |
| 06 Ara | 36 Eridanus | 66 Pisces |
| 07 Aries | 37 Fornax | 67 Piscis Austrinus |
| 08 Auriga | 38 Gemini | 68 Puppis |
| 09 Boötes | 39 Grus | 69 Pyxis |
| 10 Caelum | 40 Hercules | 70 Reticulum |
| 11 Camelopardalis | 41 Horologium | 71 Sagitta |
| 12 Cancer | 42 Hydra | 72 Sagittarius |
| 13 Canes Venatici | 43 Hydrus | 73 Scorpius |
| 14 Canis Major | 44 Indus | 74 Sculptor |
| 15 Canis Minor | 45 Lacerta | 75 Scutum |
| 16 Capricornus | 46 Leo | 76 Serpens |
| 17 Carina | 47 Leo Minor | 77 Sextans |
| 18 Cassiopeia | 48 Lepus | 78 Taurus |
| 19 Centaurus | 49 Libra | 79 Telescopium |
| 20 Cepheus | 50 Lupus | 80 Triangulum |
| 21 Cetus | 51 Lynx | 81 Triangulum Australe |
| 22 Chamaeleon | 52 Lyra | 82 Tucana |
| 23 Circinus | 53 Mensa | 83 Ursa Major |
| 24 Columba | 54 Microscopium | 84 Ursa Minor |
| 25 Coma Berenices | 55 Monoceros | 85 Vela |
| 26 Corona Australis | 56 Musca | 86 Virgo |
| 27 Corona Borealis | 57 Norma | 87 Volans |
| 28 Corvus | 58 Octans | 88 Vulpecula |
| 29 Crater | 59 Ophiuchus | |
| 30 Crux | 60 Orion | |

## 10.14 More precise pointing

### 10.14.1 How to sync on an object

If you suspect that the telescope is not well aligned with the celestial sphere, an easy way to improve the alignment is to *sync* (synchronize) with a known object.

**Table 10.2** *GCVS star numbers*

The full GCVS number consists of the constellation code (previous table) followed by the star number.

| | | | | | | | | | | | | | |
|---|---|---|---|---|---|---|---|---|---|---|---|---|---|
| 0001 | R | 0050 | XY | 0099 | BV | 0148 | DZ | 0197 | GO | 0246 | KM | 0295 | NT |
| 0002 | S | 0051 | XZ | 0100 | BW | 0149 | EE | 0198 | GP | 0247 | KN | 0296 | NU |
| 0003 | T | 0052 | YY | 0101 | BX | 0150 | EF | 0199 | GQ | 0248 | KO | 0297 | NV |
| 0004 | U | 0053 | YZ | 0102 | BY | 0151 | EG | 0200 | GR | 0249 | KP | 0298 | NW |
| 0005 | V | 0054 | ZZ | 0103 | BZ | 0152 | EH | 0201 | GS | 0250 | KQ | 0299 | NX |
| 0006 | W | 0055 | AA | 0104 | CC | 0153 | EI | 0202 | GT | 0251 | KR | 0300 | NY |
| 0007 | X | 0056 | AB | 0105 | CD | 0154 | EK | 0203 | GU | 0252 | KS | 0301 | NZ |
| 0008 | Y | 0057 | AC | 0106 | CE | 0155 | EL | 0204 | GV | 0253 | KT | 0302 | OO |
| 0009 | Z | 0058 | AD | 0107 | CF | 0156 | EM | 0205 | GW | 0254 | KU | 0303 | OP |
| 0010 | RR | 0059 | AE | 0108 | CG | 0157 | EN | 0206 | GX | 0255 | KV | 0304 | OQ |
| 0011 | RS | 0060 | AF | 0109 | CH | 0158 | EO | 0207 | GY | 0256 | KW | 0305 | OR |
| 0012 | RT | 0061 | AG | 0110 | CI | 0159 | EP | 0208 | GZ | 0257 | KX | 0306 | OS |
| 0013 | RU | 0062 | AH | 0111 | CK | 0160 | EQ | 0209 | HH | 0258 | KY | 0307 | OT |
| 0014 | RV | 0063 | AI | 0112 | CL | 0161 | ER | 0210 | HI | 0259 | KZ | 0308 | OU |
| 0015 | RW | 0064 | AK | 0113 | CM | 0162 | ES | 0211 | HK | 0260 | LL | 0309 | OV |
| 0016 | RX | 0065 | AL | 0114 | CN | 0163 | ET | 0212 | HL | 0261 | LM | 0310 | OW |
| 0017 | RY | 0066 | AM | 0115 | CO | 0164 | EU | 0213 | HM | 0262 | LN | 0311 | OX |
| 0018 | RZ | 0067 | AN | 0116 | CP | 0165 | EV | 0214 | HN | 0263 | LO | 0312 | OY |
| 0019 | SS | 0068 | AO | 0117 | CQ | 0166 | EW | 0215 | HO | 0264 | LP | 0313 | OZ |
| 0020 | ST | 0069 | AP | 0118 | CR | 0167 | EX | 0216 | HP | 0265 | LQ | 0314 | PP |
| 0021 | SU | 0070 | AQ | 0119 | CS | 0168 | EY | 0217 | HQ | 0266 | LR | 0315 | PQ |
| 0022 | SV | 0071 | AR | 0120 | CT | 0169 | EZ | 0218 | HR | 0267 | LS | 0316 | PR |
| 0023 | SW | 0072 | AS | 0121 | CU | 0170 | FF | 0219 | HS | 0268 | LT | 0317 | PS |
| 0024 | SX | 0073 | AT | 0122 | CV | 0171 | FG | 0220 | HT | 0269 | LU | 0318 | PT |
| 0025 | SY | 0074 | AU | 0123 | CW | 0172 | FH | 0221 | HU | 0270 | LV | 0319 | PU |
| 0026 | SZ | 0075 | AV | 0124 | CX | 0173 | FI | 0222 | HV | 0271 | LW | 0320 | PV |
| 0027 | TT | 0076 | AW | 0125 | CY | 0174 | FK | 0223 | HW | 0272 | LX | 0321 | PW |
| 0028 | TU | 0077 | AX | 0126 | CZ | 0175 | FL | 0224 | HX | 0273 | LY | 0322 | PX |
| 0029 | TV | 0078 | AY | 0127 | DD | 0176 | FM | 0225 | HY | 0274 | LZ | 0323 | PY |
| 0030 | TW | 0079 | AZ | 0128 | DE | 0177 | FN | 0226 | HZ | 0275 | MM | 0324 | PZ |
| 0031 | TX | 0080 | BB | 0129 | DF | 0178 | FO | 0227 | II | 0276 | MN | 0325 | QQ |
| 0032 | TY | 0081 | BC | 0130 | DG | 0179 | FP | 0228 | IK | 0277 | MO | 0326 | QR |
| 0033 | TZ | 0082 | BD | 0131 | DH | 0180 | FQ | 0229 | IL | 0278 | MP | 0327 | QS |
| 0034 | UU | 0083 | BE | 0132 | DI | 0181 | FR | 0230 | IM | 0279 | MQ | 0328 | QT |
| 0035 | UV | 0084 | BF | 0133 | DK | 0182 | FS | 0231 | IN | 0280 | MR | 0329 | QU |
| 0036 | UW | 0085 | BG | 0134 | DL | 0183 | FT | 0232 | IO | 0281 | MS | 0330 | QV |
| 0037 | UX | 0086 | BH | 0135 | DM | 0184 | FU | 0233 | IP | 0282 | MT | 0331 | QW |
| 0038 | UY | 0087 | BI | 0136 | DN | 0185 | FV | 0234 | IQ | 0283 | MU | 0332 | QX |
| 0039 | UZ | 0088 | BK | 0137 | DO | 0186 | FW | 0235 | IR | 0284 | MV | 0333 | QY |
| 0040 | VV | 0089 | BL | 0138 | DP | 0187 | FX | 0236 | IS | 0285 | MW | 0334 | QZ |
| 0041 | VW | 0090 | BM | 0139 | DQ | 0188 | FY | 0237 | IT | 0286 | MX | 0335 | V335 |
| 0042 | VX | 0091 | BN | 0140 | DR | 0189 | FZ | 0238 | IU | 0287 | MY | 0336 | V336 |
| 0043 | VY | 0092 | BO | 0141 | DS | 0190 | GG | 0239 | IV | 0288 | MZ | etc. | |
| 0044 | VZ | 0093 | BP | 0142 | DT | 0191 | GH | 0240 | IW | 0289 | NN | | |
| 0045 | WW | 0094 | BQ | 0143 | DU | 0192 | GI | 0241 | IX | 0290 | NO | | |
| 0046 | WX | 0095 | BR | 0144 | DV | 0193 | GK | 0242 | IY | 0291 | NP | | |
| 0047 | WY | 0096 | BS | 0145 | DW | 0194 | GL | 0243 | IZ | 0292 | NQ | | |
| 0048 | WZ | 0097 | BT | 0146 | DX | 0195 | GM | 0244 | KK | 0293 | NR | | |
| 0049 | XX | 0098 | BU | 0147 | DY | 0196 | GN | 0245 | KL | 0294 | NS | | |

Find any object using the built-in catalogues; then center it in the field and press and hold ENTER until you hear a beep. The display reads:

```
Coordinates
matched
```

and the telescope now assigns the coordinates of that object to its current position.

*Caution:* Resist the temptation to sync every time you move to a new object. Syncing on objects near the celestial pole or the zenith can increase rather than reduce the overall alignment error.

### 10.14.2 High-precision mode

If you need unusually accurate pointing, go to the TELESCOPE menu and turn HI-PRECISION on, as follows:

```
→TELESCOPE
  OBJECT MENU
```
(Press ENTER)

```
→1)SITE
  2)ALIGN
```

Scroll down to:

```
  8)BALANCE
→9)hi-precision
```

and press ENTER. The display changes to:

```
  8)BALANCE
→9)HI-PRECISION
```

with HI-PRECISION in uppercase to show that it is turned on.

Now, whenever you slew to an object, the telescope will first search its star catalogue, take you to a nearby bright star and invite you to sync on it:

```
HI-PRECISION
Searching...
```
(and then, after a short delay...)

```
Center STAR 219
then press GO TO
```

Center the star with N, S, E, and W; touch up the focus if you wish; and then press GO TO as instructed.

If for some reason you don't like that particular alignment star, two others are available. Press PREV and NEXT to cycle between them.

## 10.15 Training the Smart Drive (PEC)

Periodic-error correction (PEC) is a way of obtaining smoother tracking in equatorial mode by having the computer memorize a set of guiding corrections and play them back every time the worm gear revolves.

The LX200 has permanent PEC, which means that once the corrections have been recorded, they remain in nonvolatile memory, and the drive always starts in the same position in order to play them back correctly. That's why there is a small movement in right ascension when the telescope is first powered up; the worm gear is being placed in starting position.

The status of the PEC is displayed as a number next to SMART on the TELESCOPE menu. Initially this number is 21 600. It is a sum indicating the total amount of telescope movement needed in 8 minutes of sidereal time. Training will shift it slightly, but if it ends up a long way from 21 600, the PEC may have been trained at the wrong drive rate.

Before training the PEC, make sure the drive frequency, on the main menu (p. 164), is set to 60.1 and the slew rate is set to GUIDE.

On the main menu, under TELESCOPE, SMART, there are five choices:

LEARN to start training from scratch;
UPDATE to run a second training session and average the data with what is already recorded, for greater smoothness;
ERASE to erase all the recorded corrections;
DEC LEARN to record a series of northward or southward corrections (rarely needed);
DEC CORRECT to enable playback of the northward or southward corrections.

DEC LEARN and DEC CORRECT are needed only in unusual circumstances, such as when you're tracking a fast-moving comet or when there is a polar alignment error that you cannot correct. They pertain only to a particular object or part of the sky. The ordinary (right ascension) corrections, however, are beneficial all the time, and once the telescope has been trained, it need not be retrained for many months.

To start training, center a star on the crosshairs at high power and select LEARN. A number will appear on the keypad indicating the position of the worm gear and will slowly count down to zero. (When it reaches 5 you will hear a beep.) Shortly before it reaches zero, use the E and W buttons to center the star on the crosshairs and keep it centered for the next 8 minutes. When the countdown reaches zero again, the training is complete.

I have gotten excellent results training with a CCD autoguider, provided the CCD exposure and update period is less than one second long. Others report

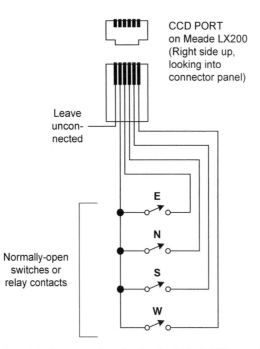

CCD PORT
on Meade LX200
(Right side up,
looking into
connector panel)

Leave uncon-
nected

Normally-open
switches or
relay contacts

E

N

S

W

Figure 10.7. External wiring for the LX200 CCD port. This is apparently equivalent to several other Meade and Celestron autoguider ports.

that manual training is better, especially if you make a box with fast-acting pushbutton switches to plug into the CCD port (Figure 10.7).

The 16-inch LX200 has permanent PEC on both axes and is pre-trained at the factory; it is not easily retrainable by the user.

## 10.16 Cables, connections, and ports

### 10.16.1 Keypad and declination cables

The modular (telephone-type) connectors on the connector panel carry power as well as control signals. Short circuits will blow fuses.

Do not connect or disconnect the keypad or declination cable while the tele-scope is powered on. The reason is that if the +12-volt line momentarily makes contact while the ground line does not, all the signal lines will go to +12 volts, causing damage.

The keypad cable is an ordinary telephone handset cable, easily replaced. The declination motor cable is a type sometimes used in the telephone industry. It has 8-pin modular plugs with the wires in the opposite order at one end than at the other. This is *not* a standard computer network cable (neither straight-through nor crossover); it is sometimes called a "rollover" cable.

One suitable replacement declination cable is Part No. H1883-05C-ND from Digi-Key Corporation, 701 Brooks Avenue South, Thief River Falls, MN 56701,

U.S.A., http://www.digikey.com. This is a silver-colored cable, easy to find if you drop it on the ground. A matching cable for the keypad is Part No. H1443-05C-ND. *Because of the risk of damage, make completely sure any replacement cable has the conductors in the right order* before installing it.

A loose declination cable will sometimes cause "runaway", a situation where the motor has power but the computer cannot sense its position, so the motor just keeps running. Note that the declination cable is joined to the modular cable coming out of the motor by a modular coupler mounted on the fork arm with one plug on either side. Either plug can come loose.

### 10.16.2 CCD port

The CCD connector allows a CCD autoguider to make guiding corrections automatically. I have also used it with a homemade box containing four pushbutton switches; when doing critical guiding or PEC training by hand, I find pushbuttons more responsive than the N, S, E, and W buttons on the keypad.

The CCD port expects to see a set of switch or relay contacts like those in Figure 10.7 or an equivalent transistor circuit. Internally, each of the inputs is pulled up to 5 volts through a resistor; the switches connect the inputs to ground.

Note that I do not give any pin numbers in Figure 10.7. The pins on the modular connector are usually numbered 1-4-2-5-3-6, which is very confusing. The diagram shows the connector as it should actually look when held with the conductors up, ready to plug into the telescope. (The unused pin is the leftmost one as you look at the connector panel.) The Meade manual shows it the other way.

When there is any activity on the CCD port, the LX200 switches to its slowest slewing rate ("Guide"), but the LED on the keypad that indicated the previous speed remains lit. This can give you the impression that the whole slewing system has gone dead. You can still select a faster slewing rate on the keypad.

### 10.16.3 The serial ports

The LX200 serial port makes it possible to control the telescope with a computer. The computer can perform all of the functions available through the keypad and can also download the LX200's object catalogues but cannot upload new data or firmware. Details and sample programs are given in the Meade manual and on http://www.mapug.com.

The LX200 actually contains *two* RS-232 serial ports and will respond to either one. This can be useful if the main serial port becomes inoperative owing to static damage, or if you devise an arcane way of using one LX200 with two software packages at the same time.

The cable available from Meade goes only to the main serial port. Figure 10.8 shows how to wire a cable of your own. This can be done without soldering if

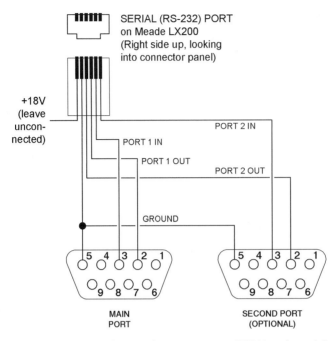

Figure 10.8. Wiring diagram for connecting an LX200 to the serial port of a PC-compatible computer. Second port is almost always unused.

Figure 10.9. Pin numbers are imprinted on serial port connectors for positive identification.

you use a solderless modular-to-DB9 adapter that lets you insert the pins where you want them.

Be sure to check the actual numbers on the pins (Figure 10.9) before trying to use the cable. Ordinarily, RS-232 connections cannot be damaged by miswiring, but one of the pins on the Meade port carries the full supply voltage (18 volts), which could damage other equipment.

### 10.16.4 Other connectors

The Aux connector supplies power for the cooling fan in the 7-inch Maksutov–Cassegrain and is reserved for other uses that will probably never be implemented.

The Focuser connector takes a miniature phone plug and outputs the full supply voltage or a reduced voltage of about 7 volts, with either polarity, for driving a focusing motor. The Meade manual describes how it is used. Neither side of the plug is grounded.

The Reticle connector supplies power to the LED in an illuminated reticle, just as on the LX3 and many subsequent Meade telescopes. Current is limited to about 7 mA so that an LED can be connected directly, without a resistor. The brightness of the LED is adjusted by holding down RET on the keypad and pressing PREV or NEXT. What actually happens is that the voltage is switched on and off about 100 times a second, and the keypad controls the fraction of the time that it is on during each cycle.

### 10.16.5 Internal battery

A button cell behind the connector panel operates the clock/calendar when the telescope is off and will require replacement every few years.

The site location and other user settings are stored in flash EEPROM and do not depend on the battery.

### 10.16.6 The floating ground

Circuit ground in the LX200, including the ground level of the RS-232 connectors, is not exactly equal to the negative side of the battery. As Figure 10.10 shows, negative battery voltage enters the LX200 through a 0.1-ohm resistor that serves

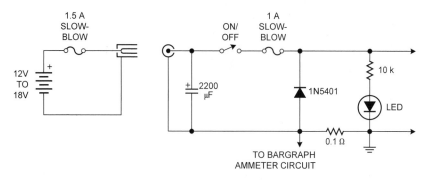

Figure 10.10. Power input circuitry of the LX200.

161

as the sensing element for the ammeter. If battery negative is tied to circuit ground externally, the LED bargraph ammeter will be inaccurate, and substantial current may flow through other equipment. The discrepancy between circuit ground and battery ground is never more than 0.2 volt.

In the circuit in Figure 10.10, the diode and internal fuse protect much of the internal circuitry against reversed polarity. If power is applied backward, the diode will conduct heavily and the fuse will blow (unless the diode burns out first). However, the 2200-microfarad capacitor is not protected and can burst, with messy results.

## 10.17 Known firmware bugs

After many revisions, firmware version 3.34L is apparently free of errors that affect telescope operation. However, the magnitude of Saturn, displayed on the keypad, is incorrect, apparently off by a factor of 10. Planetary observers will be relieved to know that Saturn is not magnitude 17.7!

The "size" of a double star, displayed on the LX200 keypad, is in tenths of arc-seconds. Thus $\epsilon^2$ Lyrae (STAR 334) is displayed as SIZE 26, i.e., 2.6″. This is apparently intentional.

Alkaid and Aldebaran are out of alphabetical order in the list of named stars.

As already noted, activity on the CCD port always switches the telescope to the slowest slewing rate ("Guide"), but the previously selected rate is still displayed on the keypad.

Some LX200s lock up if a FIELD command ("identify objects in the field", under Object Library) is issued at a right ascension between $23^h30^m$ and $0^h0^m$. Apparently the LX200 makes an arithmetic error when trying to look past the zero meridian.

## 10.18 Mechanical and electrical improvements

Many LX200 modifications are discussed on http://www.mapug.com. Two stand out as particularly useful.

The present LX200 requires a 1.5-ampere slow-blow fuse in the power cord as well as a 1-ampere fuse inside. An obvious improvement would be to replace at least one of these with a resettable circuit breaker.

The LX200 focuser is smooth but rather stiff; it uses nylon washers as bearings. Installing thrust bearings makes it much easier to turn, like a Celestron focuser. A thrust bearing is a miniature Lazy Susan; it consists of two large washers with rollers or ball bearings between them, so that each can turn easily relative to the other.

A kit for installing thrust bearings in an LX200 focuser is available from Peterson Engineering Corporation (405 New Meadow Road, Barrington, RI 02806, U.S.A., http://www.peterson-web.com). I found the installation process straightforward, though it required careful work.

You can buy an even smoother focuser ready-assembled from EZTelescope, 2713 Harper St, Lawrence, KS 66046, U.S.A. (http://www.eztelescope.com). The EZTelescope focuser has additional washers, as well as an adjustment to eliminate backlash (slack) and a massive knob that makes fine focusing easier. It is what I am currently using, and, particularly with astrophotography, it really helps.

Mechanically inclined readers may prefer to obtain their thrust bearings locally. They should be lubricated before use. Peterson sells a special lubricant; as a substitute I have used Mobil 1 Synthetic Universal Grease, which works well over a very wide range of temperatures. This grease is pinkish in color; goes onto metal surfaces in a thin layer like wax; and is unlikely to stain objects or clothing with which it comes in contact.

In 2001, Meade improved the declination bearings in the LX200 fork arms; some amateurs have installed better bearings themselves, but some mechanical skill is required. The upgrade is beneficial mainly with the 10-inch and 12-inch telescopes. More information is available on http://www.mapug.com.

## 10.19 Menu maps

The following pages show (but do not fully explain) all the menu selections available on the LX200 keypad. For more information see the Meade manual.

# Meade LX200 (v3.34)
# MAIN MENU MAP

Copyright 2000
Michael A. Covington

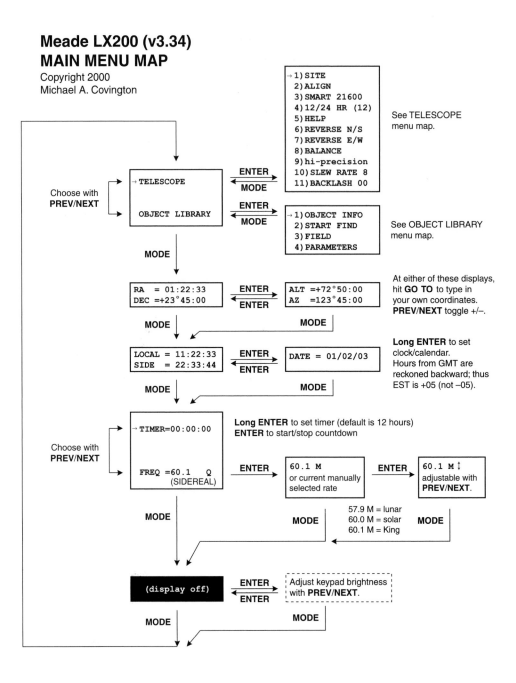

→ 1) SITE
2) ALIGN
3) SMART 21600
4) 12/24 HR (12)
5) HELP
6) REVERSE N/S
7) REVERSE E/W
8) BALANCE
9) hi-precision
10) SLEW RATE 8
11) BACKLASH 00

See TELESCOPE menu map.

→ TELESCOPE
OBJECT LIBRARY

Choose with **PREV/NEXT**

ENTER / MODE

→ 1) OBJECT INFO
2) START FIND
3) FIELD
4) PARAMETERS

See OBJECT LIBRARY menu map.

MODE

RA = 01:22:33
DEC =+23°45:00

ENTER / ENTER

ALT =+72°50:00
AZ =123°45:00

At either of these displays, hit **GO TO** to type in your own coordinates. **PREV/NEXT** toggle +/–.

MODE / MODE

LOCAL = 11:22:33
SIDE = 22:33:44

ENTER / ENTER

DATE = 01/02/03

**Long ENTER** to set clock/calendar. Hours from GMT are reckoned backward; thus EST is +05 (not –05).

MODE / MODE

→ TIMER=00:00:00

FREQ =60.1   Q
(SIDEREAL)

Choose with **PREV/NEXT**

**Long ENTER** to set timer (default is 12 hours)
**ENTER** to start/stop countdown

ENTER

60.1 M
or current manually selected rate

ENTER

60.1 M ↕
adjustable with **PREV/NEXT**.

MODE / MODE / MODE

57.9 M = lunar
60.0 M = solar
60.1 M = King

(display off)

ENTER / ENTER

Adjust keypad brightness with **PREV/NEXT**.

MODE / MODE

## Meade LX200 (v3.34)
# TELESCOPE MENU MAP
Copyright 2000
Michael A. Covington

Press
**PREV/NEXT**
to choose item.

Press **MODE**
to exit to
main menu.

→1)SITE

→ENTER→
←MODE←

| 1)AAA |
| →2)AAA |
| 3)AAA ✓ |
| 4)AAA |
| 5)UNKNOWN |

User-defined sites.
**PREV/NEXT** to choose a site,
**ENTER** to select it (indicated by ✓),
**Long ENTER** to edit site
name and location.

2)ALIGN

→ENTER→
←MODE←

| ALTAZ |
| →POLAR ✓ |
| LAND |

Alignment modes.
**PREV/NEXT** to choose a mode,
**ENTER** to select it (indicated by ✓),
**ENTER** again for alignment procedure.

3)SMART 21600

→ENTER→
←MODE←

| →1)LEARN |
| 2)UPDATE |
| 3)ERASE |
| 4)DEC LEARN |
| 5)DEC CORRECT |

Periodic-error correction.
**PREV/NEXT** to choose a selection,
**ENTER** to execute it.
R.A. programming is permanent;
dec. programming is per session.

4)12/24 HR (12)  **ENTER** to toggle 12/24-hour display of local time.

5)HELP  **ENTER** to view summary of commands, **MODE** to exit.

6)REVERSE N/S  **ENTER** to toggle. Swaps action of N and S keys.

7)REVERSE E/W  **ENTER** to toggle. Swaps action of E and W keys.

8)BALANCE  **ENTER** to start/stop balance test (alternating north-south slewing).

9)hi-precision  **ENTER** to toggle on/off. Lowercase when turned off.

10)SLEW RATE 8  **ENTER** to step through available values.

11)BACKLASH 00  **Long ENTER** to edit declination backlash compensation.

# Meade LX200 (v3.34)
# OBJECT LIBRARY
# MENU MAP
Copyright 2000
Michael A. Covington

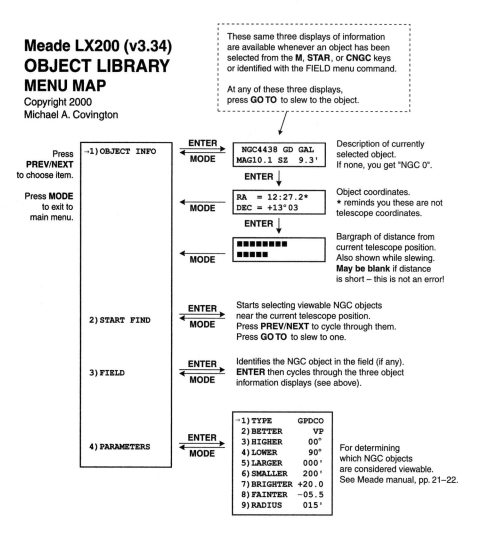

These same three displays of information are available whenever an object has been selected from the **M**, **STAR**, or **CNGC** keys or identified with the FIELD menu command.

At any of these three displays, press **GO TO** to slew to the object.

Press **PREV/NEXT** to choose item.

Press **MODE** to exit to main menu.

→1) OBJECT INFO

ENTER
MODE

```
NGC4438 GD GAL
MAG10.1 SZ  9.3'
```

Description of currently selected object.
If none, you get "NGC 0".

ENTER ↓

```
RA  = 12:27.2*
DEC = +13° 03
```

MODE

Object coordinates.
\* reminds you these are not telescope coordinates.

ENTER ↓

```
■■■■■■■■
■■■■■
```

MODE

Bargraph of distance from current telescope position. Also shown while slewing. **May be blank** if distance is short – this is not an error!

2) START FIND

ENTER
MODE

Starts selecting viewable NGC objects near the current telescope position.
Press **PREV/NEXT** to cycle through them.
Press **GO TO** to slew to one.

3) FIELD

ENTER
MODE

Identifies the NGC object in the field (if any).
**ENTER** then cycles through the three object information displays (see above).

4) PARAMETERS

ENTER
MODE

```
→1) TYPE     GPDCO
 2) BETTER      VP
 3) HIGHER     00°
 4) LOWER      90°
 5) LARGER    000'
 6) SMALLER   200'
 7) BRIGHTER +20.0
 8) FAINTER  -05.5
 9) RADIUS    015'
```

For determining which NGC objects are considered viewable. See Meade manual, pp. 21–22.

# Meade LX200 (v3.34)
## STAR MENU MAP
Copyright 2000 Michael A. Covington

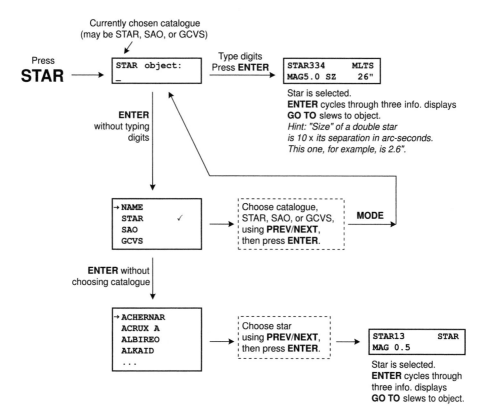

Currently chosen catalogue
(may be STAR, SAO, or GCVS)

Press
**STAR** →

```
STAR object:
_
```

Type digits
Press **ENTER** →

```
STAR334    MLTS
MAG5.0 SZ    26"
```

Star is selected.
**ENTER** cycles through three info. displays
**GO TO** slews to object.
*Hint: "Size" of a double star*
*is 10 x its separation in arc-seconds.*
*This one, for example, is 2.6".*

**ENTER**
without typing
digits

```
→ NAME
  STAR      ✓
  SAO
  GCVS
```

→ Choose catalogue,
STAR, SAO, or GCVS,
using **PREV/NEXT**,
then press **ENTER**.

**MODE**

**ENTER** without
choosing catalogue

```
→ ACHERNAR
  ACRUX A
  ALBIREO
  ALKAID
  . . .
```

→ Choose star
using **PREV/NEXT**,
then press **ENTER**.

→

```
STAR13      STAR
MAG 0.5
```

Star is selected.
**ENTER** cycles through
three info. displays
**GO TO** slews to object.

167

# Meade LX200 (v3.34)
# M, CNGC MENU MAPS
Copyright 2000 Michael A. Covington

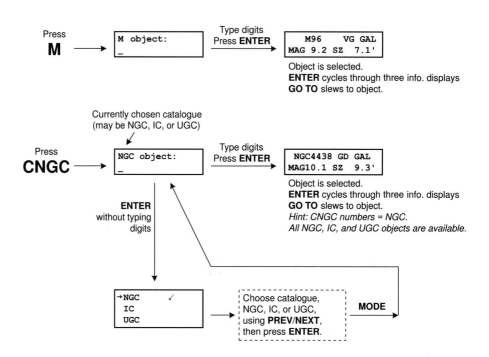

Press

**M** $\longrightarrow$

```
M object:
_
```

Type digits
Press **ENTER** $\longrightarrow$

```
 M96    VG GAL
MAG 9.2 SZ  7.1'
```

Object is selected.
**ENTER** cycles through three info. displays
**GO TO** slews to object.

Currently chosen catalogue
(may be NGC, IC, or UGC)

Press

**CNGC** $\longrightarrow$

```
NGC object:
_
```

Type digits
Press **ENTER** $\longrightarrow$

```
NGC4438 GD GAL
MAG10.1 SZ  9.3'
```

Object is selected.
**ENTER** cycles through three info. displays
**GO TO** slews to object.
*Hint: CNGC numbers = NGC.*
*All NGC, IC, and UGC objects are available.*

**ENTER**
without typing
digits

```
→NGC      ✓
  IC
  UGC
```

$\longrightarrow$

Choose catalogue,
NGC, IC, or UGC,
using **PREV/NEXT**,
then press **ENTER**.

**MODE**

168

# Chapter 11
# Celestron NexStar 5 and 8

## 11.1 Introduction

Although it came on the market slightly later than the Meade Autostar, the original model Celestron NexStar is a simpler design, so it is logical to describe it first.

This chapter is based on my experiences with a NexStar 5 purchased in 2001. This is a 5-inch (12.5-cm) Schmidt–Cassegrain and was initially the flagship of the NexStar fleet. It is practically identical to the original (one-armed) NexStar 8.

### 11.1.1 Related products

Celestron has subsequently introduced a wide range of NexStar telescopes, ranging from small refractors to an 11-inch Schmidt–Cassegrain. Some of the smaller NexStars have been sold under the **Tasco Starguide** label.

No two NexStar models have entirely the same firmware, and the product line is still evolving rapidly. Accordingly, this description of the NexStar 5 will serve only as a general guide to the rest of the product line; changes are ongoing. Many different versions of the NexStar 5 firmware are already extant.

The top-of-the-line NexStar GPS telescopes are radically different and are not covered here. They include GPS receivers to determine the latitude, longitude, and time; periodic-error correction; and other advanced features.

### 11.1.2 Evaluation of the NexStar 5

Compared with Meade LX200 and Autostar telescopes, the NexStar 5 is appreciably easier to use because it has fewer features; the overall design of the firmware is simpler. The optics are from the classic (and, in my opinion, underrated) Celestron 5. Although not much heavier than an ETX-90, this Schmidt–Cassegrain gathers twice as much light and has 40% more resolving power.

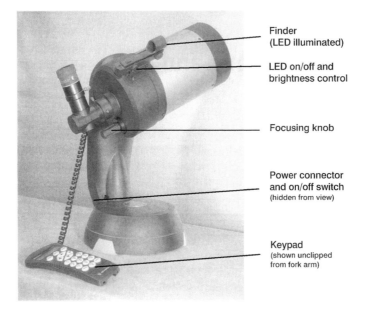

Finder
(LED illuminated)

LED on/off and
brightness control

Focusing knob

Power connector
and on/off switch
(hidden from view)

Keypad
(shown unclipped
from fork arm)

Figure 11.1. Important parts of the NexStar 5. Keypad hangs on the fork arm for storage.

The NexStar keypad is the most user-friendly that I've seen; clear, legible orange illumination is combined with a menu system that puts all important functions within easy reach.

Minor annoyances of the NexStar 5 are the fact that updates cannot be downloaded (so you're stuck with whatever firmware errors may be present) and the star catalogue does not use SAO or HIP numbers.

### 11.1.3 Firmware versions

On most NexStars, the firmware version cannot be displayed; it can still be determined by disassembling the telescope and looking for labels stuck onto four different ICs.

Early NexStars display their firmware version when powered on; some later versions display it only if all four arrow buttons (▲▼◀▶) are held down while power is turned on.

The last version number that could be displayed in this way is reportedly 4.13.13.10. The four numeric fields describe, respectively, the main ROM, the altitude microcontroller, the azimuth microcontroller, and the hand box (keypad).

### 11.1.4 NexStar websites

All the NexStar instruction manuals are available online at Celestron's website, http://www.celestron.com.

The main Internet resource for NexStar owners is a set of NexStar-related discussion groups on http://www.yahoogroups.com. There are also a number of amateur websites devoted to the NexStar. The best approach to the Web is probably to go to a major search engine, such as http://www.google.com, and type "Nexstar". In preparing these notes I have benefited greatly from Aaron Wallace's NexStar website at http://www.koessel.com/nexstar/ (apparently no longer online).

## 11.2 Important precautions

*Do not move the tube by hand.* It is designed to move only under electrical power. When necessary, you can use gentle hand pressure to move the telescope in altitude (declination), but do not try to revolve it around the base. If the altitude clutch is too loose or too tight, you can adjust it by removing the plastic cover of the fork arm and tightening or loosening the large nut inside.

*Do not screw the mounting screws too far into the base of the telescope.* The NexStar 4 and 5 can be severely damaged by screws that extend too far into the base because an internal revolving mechanism will bump into them. The maximum safe depth is only 9 mm ($\frac{3}{8}$ inch), the same as the diameter of the screws.

## 11.3 Electrical requirements

The power cord is the only thing you have to connect in order to set up the NexStar 5, since the keypad is permanently attached and the declination motor wiring is internal.

The NexStar 5 is powered by eight 1.5-volt AA cells or an external 8- to 18-volt DC power supply capable of delivering at least 750 mA. The actual current consumption, at 12 volts, is about 750 mA slewing and 250 mA tracking or idle.

Do not leave dead batteries in the telescope after use. They can leak, causing damage. Also, there have been reports of poor connections with internal batteries, causing erratic behavior. On the other hand, batteries can serve as a backup in case the external power connection comes loose while the telescope is in use. Be careful not to put a battery in backward; that's easy to do in the dark.

The power connector is a coaxial plug, 5.5 mm o.d., 2.1 mm i.d., center positive, the same as for many other Celestron telescopes.

*Caution:* The 5.5 × 2.5-mm coaxial plug used on Meade telescopes will plug into the NexStar but will not make a good connection; performance will be erratic. The same is true of 5.0 × 2.1-mm plugs that are often encountered. In fact, plugs of the latter size are even found on some Celestron AC adapters.

If you have some difficulty getting a good connection, try spreading apart the two tines of the central connector in the socket.

*Caution:* SBIG CCD cameras use a coaxial plug like that on Meade telescopes, but with the opposite polarity. If connected to a NexStar telescope, it will not work and may cause damage.

The power plug goes into the revolving base of the NexStar and has some tendency to pull out as the telescope moves. A simple precaution is to stick a

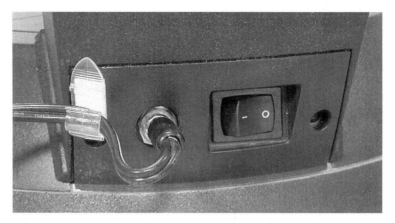

Figure 11.2. A stick-on telephone-wire clip helps keep the power cord from pulling out as the telescope rotates.

telephone-wire clip near the socket (Figure 11.2) and loop the wire through it for safety.

## 11.4  Keypad

Figure 11.3 shows the NexStar 5 keypad.

UNDO is, as its name implies, the "undo" or "back up" key. Press it to back out of a menu without making a choice, or when you get lost in the menu system. It is also the backspace key when typing numbers.

Most of the keys are recognized only when the NexStar Ready prompt is being displayed. To get to this prompt when the display shows something else, press UNDO repeatedly.

Do not confuse ▲ and ▼ (the up and down arrows, for slewing the telescope) with the UP and DOWN keys for selecting menu choices.

The RATE key works at any time, regardless of what operation is in progress, to change the slewing speed.

Messages longer than 16 characters appear as horizontally scrolling (moving) text.

The keypad illumination can be turned off (see p. 191) but the brightness is not adjustable. When very cold, the LCD display may fail to display scrolling text readably.

### 11.4.1 Direction of movement

At slew rate 6 and lower, the arrow keys (▲ ▼ ◄ ►) match the direction in which the object appears to move in the field, as seen through the telescope with the diagonal in place. For example, if an object is at the lower left, press ▲ and ► to center it.

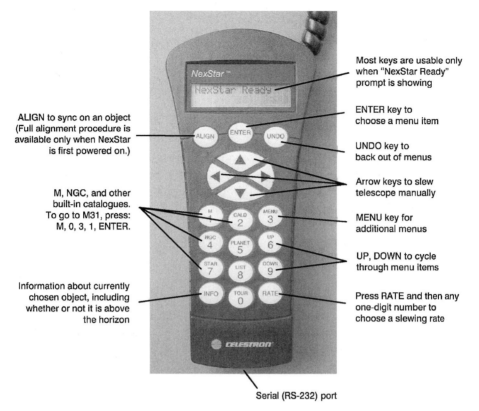

Most keys are usable only
when "NexStar Ready"
prompt is showing

ENTER key to
choose a menu item

UNDO key to
back out of menus

Arrow keys to slew
telescope manually

MENU key for
additional menus

UP, DOWN to cycle
through menu items

Press RATE and then any
one-digit number to
choose a slewing rate

ALIGN to sync on an object
(Full alignment procedure is
available only when NexStar
is first powered on.)

M, NGC, and other
built-in catalogues.
To go to M31, press:
M, 0, 3, 1, ENTER.

Information about currently
chosen object, including
whether or not it is above
the horizon

Serial (RS-232) port

Figure 11.3. The NexStar 5 keypad.

When slew rates 7, 8, and 9 are selected, the left and right arrows are swapped to match the view through the finder rather than the main telescope.

### 11.4.2 How to enter declinations and latitudes

Since a declination or latitude can be positive (north) or negative (south), entering a declination or latitude on the NexStar is a two-step process.

Initially the cursor is on the sign (+ or −), which you can change by pressing UP or DOWN. Press ENTER to accept the sign, then type the digits. Use UNDO as the backspace key.

## 11.5 Basic operation without alignment

### 11.5.1 Operation without electricity

The NexStar 5 requires electricity either from batteries or from an external source of 12 volts; there are no manual movements.

### 11.5.2 Motorized operation without alignment

To view land objects or get acquainted with the NexStar 5 without aligning on the stars, simply turn it on. When the display shows:

```
NexStar Ready
Press ENTER to begin alignment, UNDO to bypass alignment
```

(with the second line scrolling, of course), press UNDO instead. Now the telescope can slew up, down, and sideways under the control of the arrow keys, but it will not attempt to track the sky. You can adjust backlash (p. 191) and perform other tests on the instrument.

If you later want to align the telescope on the stars, you must turn it off and on again. There is no way to enter the alignment sequence from the menus.

### 11.5.3 Controlling the slewing speed

To change the speed at which the NexStar slews, press RATE and then a one-digit number. Table 11.1 shows the speeds available. "Sidereal" refers to the rate of diurnal motion at the celestial equator, i.e., the speed of a conventional (noncomputerized) telescope drive.

Note that the up and down arrows are reversed at speeds 7, 8, and 9, since it is assumed that you are looking through the finder rather than the telescope.

Regardless of the rate currently selected, you can move the telescope at maximum speed by pressing an arrow and then, while holding it, pressing the opposite arrow. This is a handy feature carried over from Celestron Celestar telescopes.

Table 11.1 *NexStar 5 slewing rates*

| Rate | Slewing speed | |
| --- | --- | --- |
| 9 | 6.5° per second* | |
| 8 | 3° per second* | |
| 7 | 1.5° per second* | |
| 6 | 32′ per second | (128 × sidereal) |
| 5 | 16′ per second | (64 × sidereal) |
| 4 | 4′ per second | (16 × sidereal) |
| 3 | 2′ per second | (8 × sidereal) |
| 2 | 30″ per second | (2 × sidereal) |
| 1 | 15″ per second | (1 × sidereal) |

* At rates 7, 8, and 9, ▲ and ▼ are reversed to match the view through the finder.

## 11.6 Entering date, time, and site information

### 11.6.1 Setting the date and time

The NexStar asks for the date and time only when it actually needs them. Thus, you will be prompted for the date and time when you initiate auto-alignment, when you ask the telescope to find a planet, or when you press MENU and choose Date & Time (p. 191).

There is no internal clock/calendar, and even the time zone is not remembered from one session to the next. The date prompt looks like this:

```
Date mm/dd/yy
00/00/00
```

Type the numbers (using UNDO as backspace if needed) and press ENTER. Be sure to enter the date American style (month, day, and year, each of them two digits). For example, 2003 September 14 is 09/14/03.

This is followed immediately by the time prompt:

```
Local Time
00:00
```

Accuracy to one minute is more than adequate. If the hour that you enter is between 1 and 12 inclusive, you will get another prompt:

```
Select One
PM            ↕
```

Choose p.m. or a.m. by pressing UP and DOWN, then press ENTER. Then choose daylight saving time (summer time) or standard time:

```
Select One
Standard Time   ↕
Daylight Saving
```

Only two lines of this display are visible at a time, of course; the second line changes as you press UP and DOWN. Again, press ENTER to make your choice.

Finally you're prompted for the time zone:

```
Select Time Zone
Pacific US Zone ↕
```

There are many other choices, including Universal Time (GMT) and some zones identified only by number, counting *westward* from Greenwich. Thus British Summer Time, GMT+1, is zone 23 on the NexStar.

Regardless of your location, you can use any time zone so long as the time and date that you enter are correct for that zone. For example, you can use Universal Time (GMT) year-round, bearing in mind that the GMT date advances to the next day at GMT midnight, early in the North American evening.

### 11.6.2 Entering site latitude and longitude

The NexStar prompts you for your location only when it needs the information, i.e., when initiating auto-alignment. You can either enter your latitude and longitude or choose a site previously stored. The first prompt that you see is:

```
Select Location
Enter Long-Lat  ↕
User Defined
```

If you take the first choice, you are prompted as follows:

```
Enter Longitude
000°00'
```

and then:

```
Enter Longitude
W
                ↕
E
```

The NexStar measures longitude the same way as ordinary atlases; North American longitudes are west. For example, Atlanta is around 84°W.

Latitude is positive for north and negative for south. (North America and Europe are all positive.) The prompt looks like this:

```
Enter Latitude
±00°00'
```

First choose + or − with UP and DOWN; then press ENTER; then enter the digits and press ENTER again.

### 11.6.3 Storing an observing site

Immediately after entering a latitude and longitude, you will be asked whether to store it:

```
Save Location
No
                ↕
Yes
```

If you choose "Yes", you get to give the location an identifying number, 0 to 9:

```
Location (0-9)
0
```

From then on, you can call up the stored location by choosing User Defined at the Select Location prompt.

*Caution:* Some versions of the firmware have a bug that causes some of the locations to overwrite others. Locations 0, 4 and 8 are always usable, but if you store data in the intermediate locations (1, 2, 3, 5, 7, 9), the information will be corrupted.

## 11.7 Aligning the telescope on the sky

### 11.7.1 Altazimuth mode

**Setting up the tripod**

At the beginning of the altazimuth alignment process, the NexStar 5 should be aimed north, with its tube level, i.e., pointing at the horizon. Great precision is not needed, which is fortunate because the NexStar 5 has no bubble level on the base or the tripod, and no declination setting circle to tell you when the tube is parallel to the base. See however p. 185.

**Checking the finder**

Before proceeding further, aim the telescope at the Moon or a very distant land object and adjust the finder to agree with the main telescope. Once set, the finder will remain properly adjusted for weeks or months.

The LED in the finder is controlled by a knob that incorporates a switch. The knob is a brightness control, and minimum brightness is almost always sufficient. Remember to turn the LED off when you've finished using it.

Many users find that the finder LED is too bright even at the minimum setting. To fix the problem, you can replace the 3-volt lithium cell with a 1.5-volt button cell surrounded by a rubber washer of the appropriate size. (Although rated at 1.8 volts, a red LED begins to glow dimly at 1.48 V.) Another tactic is to insert a 2700-ohm resistor or change the built-in rheostat to a higher value.

**Manual selection of two stars**

If you can identify two alignment stars yourself, you can set up your NexStar without entering the location, date, or time. The stars are chosen from the list on p. 183. See p. 28 for advice on which stars to choose.

Turn the NexStar on. The display shows:

```
NexStar Ready
Press ENTER to begin alignment, UNDO to bypass alignment
```

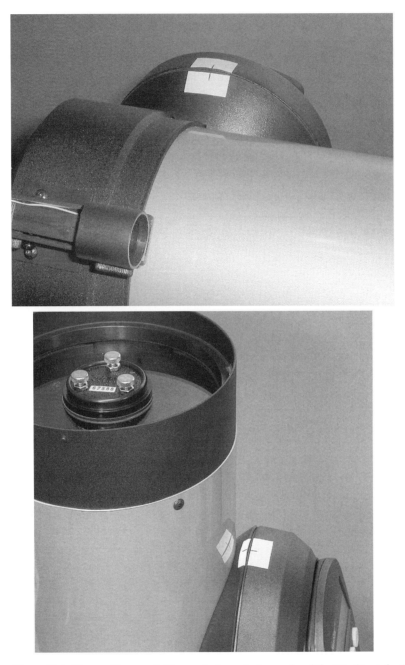

Figure 11.4. Pieces of tape, suitably marked, can help you level the NexStar tube (top) or point it away from its base for polar alignment (bottom). Exact positions can be found with a bubble level, then marked.

Press ENTER. Then, at the menu:

```
Select Method
Auto-Align     ↕
2-Star Align
```

press DOWN to select 2-Star Align, then press ENTER. The next prompt is:

```
Level the Tube
Use Direction-Buttons to move tube to approx horizontal
```

Put the tube in the starting position if you have not already done so. (Aiming it north is not actually necessary.) Then press ENTER.

Next you get a menu of stars:

```
Select Star 1
Achernar       ↕
Acrux
⋮
```

Choose two stars about 60° to 120° apart (see p. 28). Polaris is often a good choice with NexStars, even though it's not recommended with Meade telescopes.

Select the first of your stars on this menu and press ENTER. Suppose you've chosen Polaris. Then follow the instructions:

```
Center Polaris
Press Align
```

Slew the telescope to the star, changing speeds as often as you need to, and making your final approach with ▲ and ▶ in order to take up slack in the gears. Press ALIGN. Then go through the same process again:

```
Select Star 2
Achernar       ↕
Acrux
⋮
```

Choose your second star (we'll suppose it's Sirius) and press ENTER.

```
Center Sirius
Press Align
```

Again, slew to the star, center it, and press ALIGN. If all goes well, you'll get the message:

```
Align Successful
Turn Pointer Off
```

reminding you to turn off the LED in the finder.

However, all does not always go well. If the telescope says

```
Bad Alignment
```

the stars were the wrong distance apart, indicating that one of them was mis-identified. Slew the telescope back to the starting position, turn it off, turn it on again, and start over.

### Automatic star selection

If it knows your location and the date and time, the NexStar can choose the alignment stars for itself. This is actually the only reason the NexStar 5 ever needs to know your location.

Turn the telescope on. At the menu:

```
Select Method
Auto-Align        ↕
2-Star Align
```

just press ENTER to initiate auto-alignment. The next step is to put the tube in starting position:

```
North and Level
Use Direction-Buttons to point telescope North...
```

If you do not know the sky, you should be quite careful about pointing the telescope north and level, so that it points as close as possible to the stars that it selects.

When you press ENTER, the NexStar will ask for your date, time and location as described on pp. 175–177. Then the telescope will choose the first alignment star, announce something like:

```
Slewing to Vega
```

move to it, and instruct you:

```
Aim to Vega
Use Direction-Buttons...Press ENTER when ready...
```

At this point all you have to do is get the selected star into the finder and press ENTER. If the star is behind a tree or otherwise unsuitable, press UNDO to try a different one. Next:

```
Center Vega
Press Align
```

Center the star precisely, making your final approach with ▲ and ▶ to take up slack in the gears, and press ALIGN.

Then the telescope will choose another star and do the same thing. When you get it centered, it will report

```
Align Successful
```

and start tracking; the motor will start to hum.

If you get a Bad Alignment, try again. Sometimes alignments fail for no easily discernible reason. Also check for misidentified stars. In particular, Castor and Pollux are both on the list of alignment stars, but they are only 4.5° apart and look very much alike. Even experienced observers sometimes confuse them.

**Rough alignment without sighting stars**

In order to align in the daytime, you can do a *very rough* "zero-star" alignment by going through the Auto-Align procedure but without actually sighting on the stars. Just tell the telescope the stars are centered without looking at them.

In this case the pointing accuracy of the telescope is totally at the mercy of the starting position (north and level). It can be improved greatly by sighting on one or more first-magnitude stars, visible in the telescope even in daylight provided the sky is clear and the telescope is already focused on infinity.

### 11.7.2 Equatorial mode

**Two-star alignment**

To use the NexStar 5 on an equatorial wedge, set up the wedge in the normal way (p. 43) and do a two-star alignment, just as in altazimuth mode. That's right – you don't have to tell the NexStar that it's on a wedge. Just select two stars and align on them. Pointing accuracy will be very good whether or not the polar alignment is accurate. There is no specified starting position, but that shown in Figure 11.5 works well.

For smoother tracking, you should switch into equatorial tracking mode after alignment. Press MENU and choose Mode, then EQ North (assuming you're in the northern hemisphere). The motor will immediately become much quieter. The telescope, if it's a later version, will also stop balking at locations that it falsely thinks are below the horizon. You can also choose the tracking rate: sidereal, King (like sidereal but correcting slightly for atmospheric refraction), solar, or lunar.

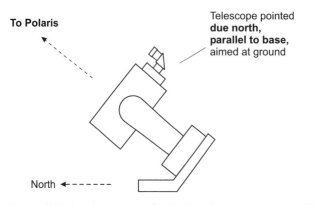

Figure 11.5. Starting position for NexStar 4 or 5 on an equatorial wedge.

### Polar-aligning the NexStar

The NexStar 5 has no declination setting circle to tell you when it's pointed directly away from its base. Accordingly, sighting on Polaris for initial polar alignment is meaningless unless you either align the base with a separate telescope (Figure 4.8, p. 47) or make a mark on the NexStar to take the place of the 90° mark on the missing setting circle (Figure 11.4, p. 178).

You can refine the polar alignment by the drift method (p. 49), making *sure* to switch to equatorial tracking first. (If you stick with altazimuth tracking, the NexStar will track the stars regardless of the polar alignment error; it's too smart for its own good.) Going to Polaris or aligning iteratively (à la Meade) will not work because the NexStar does not assume that its axis is parallel to that of the Earth.

### Improvised equatorial mode on NexStar 4

The NexStar 4 offers only Auto-Align, not manual two-star alignment, and its firmware makes no provision for equatorial mode. However, you can trick it.

To do so, set the latitude to +89° (the farthest north it will go), but use your real longitude and time zone. When told to point the tube north and level, point it due north, parallel to the base, pointing toward the ground (Figure 11.5).

Then proceed with the alignment process. The telescope will think you're close to the North Pole and have a slightly nonlevel tripod. You can do piggyback astrophotography with occasional guiding corrections.

## 11.8  How to interrupt a slewing movement

If the telescope is on its way to an object and you need to stop it for some reason, perhaps to keep it from tangling cords or bumping into an obstacle, just hit any of the arrow buttons (▲▼◀▶) while it's moving. Alignment is not lost when you do this.

## 11.9 Finding objects with the built-in catalogues

### 11.9.1 Messier, Caldwell, and NGC objects

The NexStar 5's internal catalogues include all 109 Messier objects (numbered 1 to 110 with one duplication), all 109 Caldwell objects, and all 7840 NGC objects.

To go to Messier 31, press M, and at the prompt:

```
Messier (1-110)
000
```

type in 031, using UNDO for backspace if needed. You must enter three digits. Then press INFO for information about the object, or ENTER to slew to it. See the menu map on p. 189.

Note that the NexStar will happily slew to objects that are below the horizon![1] That's why it's a good idea to press INFO first and make sure the altitude isn't negative. The NexStar 5 does not bump into anything when slewing to a negative altitude, but a large dewcap or piggyback camera mount might run into problems.

The CALD(well) and NGC buttons work the same way. Caldwell numbers are always three digits; NGC numbers are always four digits. Thus Caldwell 1 goes in as 001, and NGC 1 goes in as 0001.

### 11.9.2 Stars

The 40 named alignment stars recognized by the NexStar 5 are:

| | | | | |
|---|---|---|---|---|
| Achernar | Alpheratz | Deneb | Mizar | Scheat |
| Acrux | Altair | Denebola | Navi | Sirius |
| Albireo | Antares | Dubhe | Peacock | Spica |
| Aldebaran | Arcturus | Fomalhaut | Polaris | Suhail |
| Algenib | Betelgeuse | Hadar | Pollux | Vega |
| Alpha Centauri | Canopus | Hamal | Procyon | |
| (Rigel Kent) | Capella | Mimosa | Rasalhague | |
| Alphard | Caph | Mirach | Regulus | |
| Alphecca | Castor | Mirfak | Rigel | |

You can select any of these by pressing LIST and choosing Named Star.

The NexStar 5 also has a built-in catalogue of 10 385 stars down to magnitude 7, accessed through the STAR button. Unfortunately, the NexStar uses its own numbering system rather than that of a standard catalogue. The complete

---

[1] At least, the original model will. A later version of the firmware refuses to do so unless you press ENTER a second time.

catalogue, with cross-indexing to SAO numbers, is available from http://www.celestron.com, on the NexStar 5 page.

### 11.9.3 Planets

The Sun, Moon, and planets are accessed through the PLANET button. As with deep-sky objects, you can press INFO to find out whether the object is above the horizon.

### 11.9.4 Lists of objects

Somewhat arbitrary lists of double stars, variable stars, asterisms, and "named objects" (well-known nebulae, clusters, and galaxies) are accessible through the LIST button and its submenus (p. 190).

### 11.9.5 Sky tours

The NexStar can even give you a tour of the sky. Press TOUR and choose the current month. The NexStar will construct a customized menu of interesting objects that are (probably) high in the sky. Choose one and press INFO to find out what it is; then press ENTER to slew to it or UNDO to skip it. In either case the menu will reappear on the display so that you can choose another object.

## 11.10 Finding objects by coordinates

### 11.10.1 Slewing to a given R.A. and declination

To go to a specified right ascension and declination, press MENU, choose Goto RA-DEC, and enter the coordinates. (See the note on entering declinations on p. 173.)

Right ascension is given in hours and decimal minutes (e.g., $12^h31.6^m$). To convert from hours, minutes, and seconds, see the chart on p. 149.

### 11.10.2 Slewing to a given altitude and azimuth

To go to a specified altitude and azimuth, press MENU, choose Goto Alt-Az, and enter the coordinates. Altitudes are always positive, but you must press ENTER to accept the plus sign before entering the digits.

The NexStar measures azimuth in the normal way; north is $0°$, east is $90°$, and so on to $360°$.

### 11.10.3 The "User Object" catalogue

You can store up to 25 objects in the NexStar's nonvolatile memory, where they remain until erased. Objects 00 to 20 are located by right ascension and declination, and objects 21 to 25 are located by altitude and azimuth.

See the menu map on p. 192. Objects can be stored either by entering their coordinates or by pointing the telescope at them and recording its position.

## 11.11 More precise pointing

### 11.11.1 Approaching with ▲ and ▶

The NexStar 5 always finishes slewing to an object in the direction corresponding to the direction buttons ▲ and ▶ (up and right). For best results, you should use the same two keys when centering an alignment star. Use slewing rate 6 or below, of course, so that the up and down arrows are not reversed.

### 11.11.2 Backlash adjustment

Backlash is slack in the gears, evident when you slew in the opposite direction than the previous slewing movement. To judge it, slew back and forth, first up and down and then left and right, while viewing an object at medium power, using the slew rate that is most important to you (typically 5).

The NexStar has electronic backlash compensation which you can adjust via the MENU button (see p. 191). If the compensation is too low, there is a long delay before the telescope starts moving; if too high, the movement starts with a sudden jerk.

Always err on the low side; delays are better than jerks. You will probably need different backlash settings in equatorial than in altazimuth mode. When guiding photographs, I generally set the backlash compensation to zero, since delays are tolerable but jerks are not.

The NexStar 5 and 8 remember the backlash settings; the NexStar 4 and below require adjusting every time you use them.

### 11.11.3 How to sync on an object

To sync on an object, center it in the telescope and press ALIGN. You will be asked which of your two alignment stars you want to replace. Choose one that is near the object, so that the other star will remain in the computer's memory as a reference point farther away. You do not want to end up with two alignment stars that are close together.

### 11.11.4 The controversy over tripod leveling

Some NexStar 4 and 5 owners report that pointing accuracy depends crucially on starting the telescope in the right position (north and level), with a level tripod. Others report that the only real requirement is that the tube must start out perpendicular (orthogonal) to the base.

My experience, however, is that large errors in the starting position and tripod leveling have no effect once two stars have been sighted. It is possible that some versions have a firmware error that causes the initial position to affect things that it should not. If you have a NexStar, some experimentation is in order.

## 11.12 Cables, connections, and ports

### 11.12.1 Keypad cable

The NexStar 5 keypad cable is permanently attached to the keypad and not easily replaceable by the user. The connection to the telescope is a six-pin modular plug, compatible with a standard six-conductor telephone extension cord. Be forewarned that once you unplug the keypad, considerable dexterity is required to get the plug back in!

### 11.12.2 Serial (RS-232) port

Through the RS-232 port, the NexStar can be controlled by a computer; it accepts commands to go to particular coordinates (R.A. and declination or altitude and azimuth) and to report its current position. That doesn't sound like much, but it's enough. Details are in the NexStar manual.

The serial cable plugs into the keypad, not the base, and uses the same kind of plug as a telephone handset cable (smaller than a regular telephone jack). Its wiring is shown in Figure 11.6. Communication is at 9600 baud, 8 data bits, no parity.

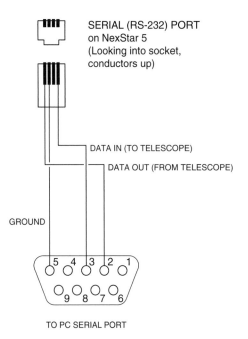

Figure 11.6. Serial cable for NexStar 5. On pin numbering see also Figure 10.9, p. 160.

The RS-232 port works only when you have pressed MENU and chosen RS-232. In RS-232 mode you can still use the direction buttons and RATE button. To escape from RS-232 mode, press UNDO.

## 11.13 Known firmware bugs

There have been many versions of the NexStar firmware, all of which have occasional but infrequent problems with pointing and slewing. My NexStar 5 occasionally goes to the wrong part of the sky for unknown reasons; sending it to the same object again results in correct performance.

On my NexStar, the positions of M2 and M10 are incorrect in the catalogue; some users report that M110 is also incorrect. M10 can be found by its NGC number (NGC 6254), but NGC 7089, which corresponds to M2, shares M2's error. When in doubt, find the object by its right ascension and declination.

Early versions of the NexStar 4 reportedly had a firmware bug that caused seriously inaccurate tracking (but not pointing). A fix is available from Celestron.

## 11.14 Menu maps

The following diagrams are a more or less complete map of the NexStar 5 menus. Other models of NexStar are similar but may have different features.

## NexStar 5 (early 2001)
# POWER-ON SEQUENCE
Copyright 2001
Michael A. Covington

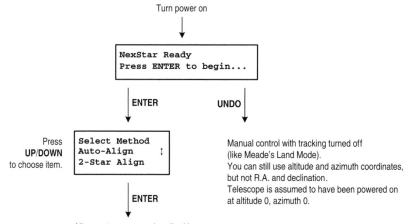

Turn power on

> NexStar Ready
> Press ENTER to begin...

ENTER · UNDO

Press UP/DOWN to choose item.

> Select Method
> Auto-Align        ↕
> 2-Star Align

Manual control with tracking turned off (like Meade's Land Mode).
You can still use altitude and azimuth coordinates, but not R.A. and declination.
Telescope is assumed to have been powered on at altitude 0, azimuth 0.

ENTER

Alignment process as described in text.
*Note:* Alignment can only be initiated by turning the telescope off and on again.
The **ALIGN** key only allows you to replace a star in the existing alignment.

## NexStar 5 (early 2001)
# M (MESSIER) MENU MAP
Copyright 2001
Michael A. Covington

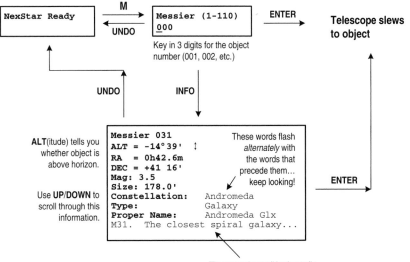

**CALD, NGC, STAR** menus work similarly except that:

NGC numbers have 4 digits (0001-7840);

STAR numbers have 5 digits and refer to the NexStar's own catalogue.

**NexStar 5** (early 2001)
## LIST MENU MAP
Copyright 2001
Michael A. Covington

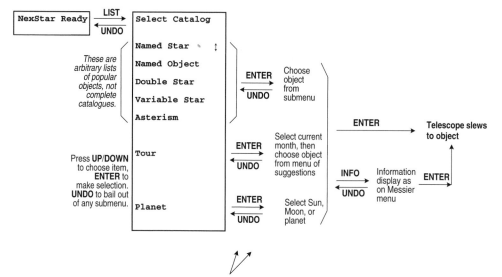

TOUR and PLANET keys on the keypad go
directly to the "Tour" and "Planet" submenus.

## NexStar 5 (early 2001)
# "MENU" MENU MAP
Copyright 2001
Michael A. Covington

```
NexStar Ready
```

MENU ↓    ↑ UNDO

Press
**UP/DOWN**
to choose item,
**ENTER** to
make selection.
**UNDO** to bail out
of any submenu.

| Menu | | |
|---|---|---|
| **Tracking** ↕ | Set tracking mode (off, equatorial in northern or southern hemisphere, or altazimuth) and rate (sidereal, lunar, solar, or King). |
| **Anti-Backlash** | Set backlash compensation to prevent jerks or delays in movement. Suggested value: 30 in altitude and azimuth. Too little is better than too much. |
| **Cordwrap** | Choose Power Cord if using a cord, Battery if not. |
| **Light Control** | Turns keypad illumination on and off. |
| **Date & Time** | Sets date, time, and time zone. NexStar will prompt for this information at any time if it is needed and you have not yet entered it. |
| **User Objects** | See **USER OBJECTS** menu map. |
| **RS-232** | Switches to serial port control. **UNDO** to exit. |
| **Demo** | Telescope slews randomly until you press a slewing button. |
| **Goto Alt-Az** | Go to a specified altitude and azimuth. |
| **Goto RA-DEC** | Go to a specified right ascension and declination. *Note:* Entering a declination is a two-step process Choose + or − with UP/DOWN, press Enter, type numbers, press Enter. |
| **Get Alt-Az** | Display altitude and azimuth at which the telescope is pointed. Assumes telescope is aligned on the stars or was powered up perfectly level and pointing north (altitude 0, azimuth 0). |
| **Get RA-DEC** | Display R.A. and declination at which the telescope is pointed. |

## NexStar 5 (early 2001)
## USER OBJECTS MENU MAP
Copyright 2001
Michael A. Covington

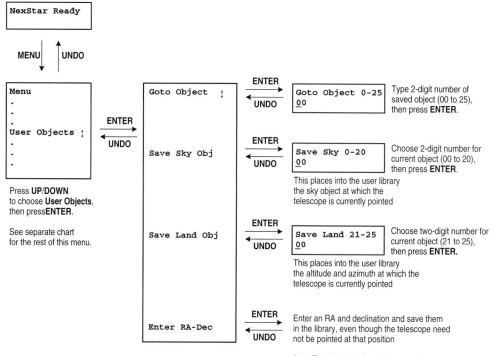

NexStar Ready

MENU ↓ ↑ UNDO

Menu
.
.
.
User Objects ↕
.
.
.

ENTER →
← UNDO

Press **UP/DOWN**
to choose **User Objects**,
then press**ENTER**.

See separate chart
for the rest of this menu.

Goto Object ↕

ENTER →
← UNDO

Goto Object 0-25
0 0

Type 2-digit number of
saved object (00 to 25),
then press **ENTER**.

Save Sky Obj

ENTER →
← UNDO

Save Sky 0-20
0 0

Choose 2-digit number for
current object (00 to 20),
then press **ENTER**.

This places into the user library
the sky object at which the
telescope is currently pointed

Save Land Obj

ENTER →
← UNDO

Save Land 21-25
0 0

Choose two-digit number for
current object (21 to 25),
then press **ENTER.**

This places into the user library
the altitude and azimuth at which the
telescope is currently pointed

Enter RA-Dec

ENTER →
← UNDO

Enter an RA and declination and save them
in the library, even though the telescope need
not be pointed at that position

*Note:*Entering a declination is a two-step process
First choose + or – with **UP/DOWN**, press **ENTER**,
type the numbers, and press **ENTER** again.

# Chapter 12
# Meade Autostar (ETX and LX90)

## 12.1 Introduction

In 1999, Meade Instruments made the sensible decision to use the same firmware, as far as possible, in all their computerized telescopes. The result was the Autostar system, used in a wide range of instruments.

The first popular Autostar telescope was the ETX-90 EC, the computerized version of a 9-cm Maksutov–Cassegrain that already had a reputation for excellent optics. The Autostar line quickly extended downward to the Meade Digital Electronic Series (DS) refractors and Newtonians, and upward to the LX90 Schmidt–Cassegrains, which are a lower-cost alternative to the LX200 for those who are not doing advanced astrophotography.

This chapter is based on my experience with an ETX-90 EC (Figure 12.1), but virtually everything in it applies to all models of Autostar. The menu system is mapped on p. 211.

Unlike some of its competitors, Autostar is relatively well documented, and I assume that you have Meade's manuals available for reference. In particular, the LX90 manual is worth reading even if you are using an ETX or one of the other lower Autostar models. By the time you read this, it should be online at http://www.meade.com. It is already online at http://www.astronomics.com.

### 12.1.1 Related products

The Autostar II system used in the LX200 GPS is an enhanced version of Autostar with periodic-error correction and a keypad that makes more of the menus available at a single keystroke.

There are several models of the original Autostar controller. The Autostar 497 works with DS, ETX, and LX90 telescopes and has a 14 000-object library. The Autostar 495, used on the DS series, has a 1200-object library.

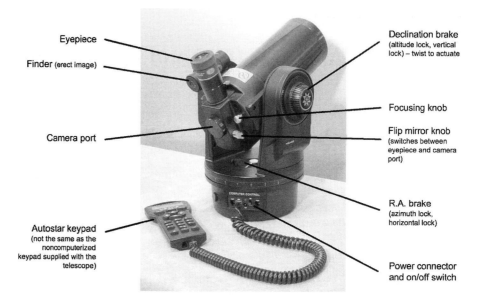

Figure 12.1. Main parts of the Meade ETX-90. Other ETX and LX90 telescopes are similar.

The bottom-of-the-line Autostar 494, supplied with the ETX-60 and ETX-70 refractors, lacks number keys; on it, numbers are selected with the ▲ and ▼ keys at the bottom of the keypad.

### 12.1.2 Evaluation of the Autostar (ETX-90 and LX90)

The Autostar firmware is much more elaborate than that of the LX200 or Celestron NexStar; besides telescope control and object catalogues, it includes an online dictionary of astronomical terms, calculators for predicting astronomical events, and other accessories, some of which will be only rarely used. Only the most important features are described here. See the menu maps (p. 211) for an indication of what else is provided.

The Autostar system requires you to type rather slowly; it does not recognize fast keystrokes. It also requires more keystrokes than its competitors for many common operations, since the menus are nested to greater depth. For example, there is no NGC key; instead, you must choose Object, Deep Sky, NGC. Both of these problems are addressed by the Autostar II.

### 12.1.3 Firmware versions

The firmware described here is version 22Er (2.2), released in late 2001. Updates can be obtained from http://www.meade.com and downloaded into the telescope through its serial port. Updating the firmware regularly is strongly recommended, but there have occasionally been buggy updates, so it is a good idea to see whether the latest upgrade has been well received by the user community.

### 12.1.4 Autostar websites

There is a strong Autostar presence on the Internet, including not only the Meade Advanced Products Users' Group (http://www.mapug.com) but also at least four ETX groups and one LX90 group on http://www.yahoogroups.com. A web search for *ETX* or *LX90* will turn up many relevant websites; since all Autostar telescopes have essentially the same firmware, information about any of them is likely to be relevant to the others. One that I have found particular helpful is Michael L. Weasner's ETX site, http://www.weasner.com/etx.

## 12.2 Electrical requirements

The ETX-90 and LX90 telescopes require 12 volts DC and can be powered by an external power supply or internal batteries. A slightly higher voltage, up to about 15 volts, gives smoother operation. The ETX-90 draws about 150 mA tracking, 500 mA slewing.

The coaxial power connector is the same as for the LX200 (p. 136) and the same cautions apply. Unlike the LX200, ETX and LX90 telescopes do not require an external fuse, but a 1.5-ampere slow-blow fuse in the power cord is still a good idea.

The ETX-60 and 70 run on 9 volts.

## 12.3 Keypad

Figure 12.2 shows the main type of Autostar keypad. Note one opportunity for confusion – two different pairs of keys are labeled ▲ ▼. The four keys at the top, ▲ ▼ ◄ ►, are the *slewing keys* or *slewing arrows*. The two keys ▲ ▼ at the bottom are for scrolling through menus and are called *scroll(ing) keys* or *scroll(ing) arrows.*

Some of the least expensive ETX telescopes come with a simpler keypad that has no digit keys; on it, you choose digits with the scrolling arrows.

The main menu begins with:

```
Select Item
 Object
```

(to which you can always return by pressing MODE repeatedly and then scrolling up or down to Object) and the rest of the menus are shown in diagrams from p. 211.

### 12.3.1 Is the computer included?

On some Meade DS and ETX telescopes, *the Autostar computer is an optional accessory.* The keypad supplied with the telescope is a noncomputerized analog

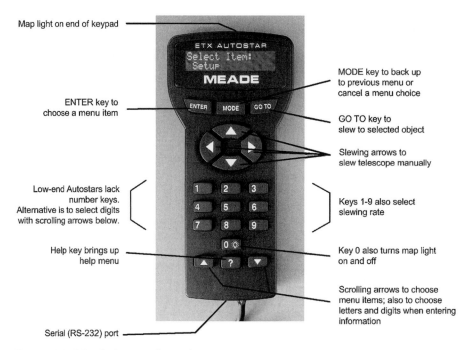

Map light on end of keypad

ENTER key to
choose a menu item

Low-end Autostars lack
number keys.
Alternative is to select digits
with scrolling arrows below.

Help key brings up
help menu

Serial (RS-232) port

MODE key to back up
to previous menu or
cancel a menu choice

GO TO key to
slew to selected object

Slewing arrows to
slew telescope manually

Keys 1-9 also select
slewing rate

Key 0 also turns map light
on and off

Scrolling arrows to choose
menu items; also to choose
letters and digits when entering
information

Figure 12.2. Meade Autostar keypad.

controller with slewing arrows and a couple of other buttons. Crucially, it does not have a GO TO key, nor is it labeled "Autostar". Do not confuse this with the low-end Autostar that is computerized but lacks number keys.

### 12.3.2 Please type slowly!

The Autostar scans its keypad more slowly than competing systems or other handheld digital devices. Not only that, but keystrokes are recognized when the key is released, not when it is pressed.

Hold down each key at least $\frac{1}{4}$ second in order for it to be recognized. A "long ENTER" (to sync on an object) or "long MODE" (to display right ascension and declination) takes at least 2 seconds. Keystrokes are recognized when the key is released, not when it is pressed.

### 12.3.3 How to enter information

When typing information into the Autostar, you can type digits directly or choose characters (A–Z, 0–9, or blank) with the scrolling keys. Use the slewing keys ◀ ▶ to move the cursor left and right.

Choose items such as names of months, North vs. South, etc., with the scrolling keys after placing the cursor on the item to be changed. Use the same technique to choose letters (A to Z) in places where they are permitted, such as site names.

Edit a plus or minus sign (+ or –) by placing the cursor on it and then pressing the scrolling key ▲ for + or ▼ for –.

### 12.3.4 Display adjustments

Many of the messages displayed by the Autostar do not fit on the screen all at once and are displayed scrolling horizontally. The speed of this scrolling can be adjusted with ▲ ▼ at the bottom of the keypad.

To adjust the display brightness, press MODE until you are at the top of the menu system:

```
Select Item
 Object
```

Then use the scrolling keys to choose Utilities, press ENTER, choose Brightness Adj., press ENTER again, use the scrolling keys to adjust the brightness, and press ENTER one last time. The selected brightness will be remembered permanently, until you change it.

Even when turned completely off, the display can be read by the light of a flashlight. It turns itself off after several minutes of disuse and can be reawakened by pressing MODE.

### 12.3.5 Direction of movement

Unless you specify otherwise, the slewing keys ▲ ▼ ◄ ► match the direction in which the telescope actually moves.

That means that when you look through the eyepiece using a diagonal or the built-in flip mirror, the slewing keys seem to be reversed from left to right but not from top to bottom. That is, ▲ ▼ move the *field* up and down, respectively, relative to the object, but ◄ ► move the *object* left and right, respectively, relative to the field.

You can program the Autostar to reverse either set of buttons to get whatever behavior you find most convenient, e.g., to match a star map that you are using.

### 12.3.6 Display modes

By holding down MODE for about 2 seconds, you can display information other than the menu system. For details see the chart on p. 217. Press MODE to get back to the menus.

## 12.4 Power-on sequence, date, and time

When first turned on, the Autostar displays its firmware version and then a warning:

```
WARNING
LOOKING AT OR NEAR THE SUN WILL CAUSE IRREVERSIBLE DAMAG...
```

That is, you shouldn't aim the telescope at or near the Sun without a full-aperture filter. Whether or not the whole message has scrolled by, you can exit from it by pressing the 5 key.[1]

Next the Autostar will offer you some instructions:

```
Getting Started
For a detailed description of the Autostar system, press...
```

You can press ? for more information, but ordinarily you'll just press ENTER. You must then enter the date and time:

```
Enter Date:
16-Sep-2001
```

If you will just be looking at land objects, you don't actually have to enter the *correct* date and time; you can just hit ENTER ENTER ENTER to accept the default values.

Correct date and time are, however, required to find sky objects. The Autostar remembers the date from the previous session but does not keep time when powered off.

Type the digits or select them with the scrolling keys (▲ ▼ at the *bottom* of the keypad). Select the month with the scroll keys. Move the cursor left and right with the slewing keys ◄ ► at the *top* of the keypad. Press ENTER and you'll be prompted for the time:

```
Enter Time:
08:00:00PM
```

Enter the digits, and select AM or PM with the scrolling keys, and press ENTER. Finally, one more question:

```
Daylight Savings
>YES
 NO
```

Here you only see two lines at a time, and > means that YES is selected. Scroll up or down until the correct value is displayed and select it by pressing ENTER.

At this point *the display suddenly goes dim* and you may think something is wrong. Actually, the display has just dimmed to the brightness that you selected for night work, and which you can adjust (p. 197). Then the Align menu comes up, but you can bail out of it by pressing MODE.

---

[1] You can set the Autostar so that it does not display this warning, nor the "Getting Started" message that follows it; see p. 216.

If you've just turned on your Autostar for the first time, you must also enter your location (see next section) and confirm your telescope model (ETX-90, LX90, etc.).

## 12.5 Entering site information

When using a new Autostar for the first time, or after erasing memory by choosing `Setup Reset`, you must enter your location, which the Autostar remembers until you change it. You can also enter your location at any time by choosing `Setup Site`.

Before proceeding I should explain that the Autostar keeps a list of up to six working sites. Initially the list is empty, and you must add at least one site to it, which you can do by picking a city from a menu or entering a latitude and longitude. To learn how to add or delete other sites, see the menu map on p. 214.

### 12.5.1 Choosing your location from a menu

The easiest way to enter your location is to select it from the menu:

```
Country/State
 AFGHANISTAN
 ALABAMA
 ALASKA
   ⋮
 CUSTOM
```

Scroll up and down until your country or state is on the screen, then press ENTER. If you want to enter your latitude and longitude, choose CUSTOM.

*Hint:* To go directly to the last item on a menu (in this case CUSTOM), scroll *up* from the first item. The menus wrap around.

Suppose you've chosen:

```
Country/State
 GEORGIA
```

When you press ENTER, you'll get a menu of cities within Georgia:

```
Nearest City
 ALBANY
 ALMA
 ATHENS
   ⋮
```

Choose any location within about 150 miles or 200 km and press ENTER. There –
you've entered your latitude, longitude, and time zone.

### 12.5.2 Entering latitude, longitude, and time zone directly

If you've chosen to enter a custom site, you'll be prompted to give it a name:

```
Site Name:
SITE
```

Choose characters (A–Z, 0–9, or blank) with the scrolling keys, and use the slewing keys ◀ ▶ to move the cursor. The name can be up to 16 characters long.
Then press ENTER and enter the latitude:

```
Enter Lat:
00°00'North
```

To change North to South, move the cursor onto it and then use the scrolling
keys. Then enter the longitude the same way:

```
Enter Lon:
000°00'East
```

(In the United States be sure to change East to West.)

Finally, enter the time zone, in hours relative to UT. Use the zone number that
is applicable when daylight saving time (summer time) is not in effect. Zones
west of Greenwich are negative; thus the U.S. Eastern zone is −5 (not +5 as it
would be on the LX200).

```
Time Zone:
-00.0
```

To change − to +, move the cursor onto it and press the scrolling keys ▲ for + or
▼ for −.

## 12.6 Basic operation without alignment

### 12.6.1 Operation without electricity

The ETX and LX90 can be used, very clumsily, without any source of power;
just unlock the R.A. and declination brakes and aim the telescope by hand.
Fine adjustments are not provided and anything more than a rough test of the
telescope is probably not practical.

## 12.6.2 Land mode

To use Autostar telescopes in "land mode", with motorized slewing but without attempting to track the stars, just skip the alignment procedure. That is, at the prompt:

```
Align
 Easy
```

press MODE to bail out. Or go to the Setup menu and choose Targets, Terrestrial.

You can now aim the telescope by pressing the slewing buttons. *It is a good idea to practice looking at land objects this way before trying the telescope on the sky.*

Note that the motors move the telescope only when the R.A. and declination brakes are locked. Note also that with the ETX-90, there is a limit to how far you can go in azimuth (left and right); the telescope covers more than 360° but will not go around and around repeatedly.

## 12.6.3 Controlling the slewing speed

Whenever the keypad is not expecting you to type a number, you can choose a slewing speed by pressing any digit from 1 (the slowest) to 9 (the fastest). The speed is displayed as you hold the digit down. For example, 64x means 64 times sidereal rate, i.e., $64 \times 15''$ per second. The speeds are mostly the same as for the NexStar (Table 11.1, p. 174), but unlike the NexStar, the Autostar does not swap any of the buttons at any of the speeds.

I generally use speed 8 or 9 to move rapidly toward an object and speed 5 to center it. The lowest speeds are for guiding during long photographic exposures.

## 12.7 Aligning the telescope on the sky

### 12.7.1 Checking the finder

Aligning on the stars is difficult unless the finder is reasonably accurate. Before setting up for astronomy, sight a distant land object and make sure the finder is aligned with the telescope.

### 12.7.2 Altazimuth mode

**Setting up the tripod**

If you will be doing a two-star alignment, tripod leveling is not critical, but the tripod head should be approximately level.

Initially, the telescope should be level (pointing at the horizon) and pointing north. Again, if you are doing a two-star alignment, this position is not highly critical.

The ETX telescope should start within 180° of its anticlockwise limit. To achieve this, unlock the R.A. brake, swivel the telescope as far to the left (anti-clockwise) as it will go (without forcing it), then to the right (clockwise) back to north.

### Two-star alignment (RECOMMENDED)

Select two bright stars as described on p. 28. In what follows I'll assume you've chosen Altair and Enif. At the prompt:

```
Align:
 Easy
```

press the lower ▼ key to scroll down to:

```
Align:
 Two Star
```

and press ENTER. The Autostar displays some instructions:

```
AltAz Align
Put the telescope in the Alt. Az. home position...
```

*Important:* If the display says Polar Align, the telescope is not in altazimuth mode. Press MODE twice to get back to the Setup menu, then choose Telescope, Mount, AltAz, and start over.

   Press ENTER and you get a menu of stars:

```
Select Star:
 ACAMAR
 ACHERNAR
 ACRUX
 ADARA
   ⋮
```

Select your first star and press ENTER. The telescope slews to the approximate location of the star. Wait until the Slewing... message goes away, then *wait for the beep* (which indicates that slewing is truly complete), and finally follow the instructions:

```
Ctr. ALTAIR
Press ENTER
```

You can change slewing speeds while doing this; in fact, you'll probably start at rate 9 and then switch to about rate 5.

The telescope will then prompt you to choose a second star and center it the same way. When you have done so, it will calculate for a moment and should then say:

```
Align Successful
```

Then it will begin tracking (with characteristic groaning noises if it's an ETX). If the alignment is *not* successful, try again, and check that you haven't misidentified either star.

After displaying this message briefly, the Autostar goes to the top of its menu system:

```
Select Item
  Object
```

and you are ready to choose objects to observe.

### Easy alignment (automatic selection of stars)

If at the start, you choose

```
Align:
  Easy
```

rather than `Align, Two Star`, the procedure is the same except that the telescope chooses the two stars for you. If one of them turns out to be behind a tree or otherwise unsuitable, press the lower ▼ key and you'll be offered another choice. You can do this as soon as the telescope starts moving.

### One-star alignment

If you choose

```
Align:
  One Star
```

the telescope will be aligned on one star and will assume that the tripod is perfectly level. One-star alignment is not recommended.

### Zero-star approximate alignment

If you cannot sight any stars, you can align the telescope *roughly* by going through a one- or two-star alignment procedure and telling it that the stars are centered even if they aren't. This is obviously a somewhat dubious method, but in some cases it's better than nothing.

Another way to shortcut the alignment process is to bail out of Setup, Align by pressing MODE, then slew to any celestial object and sync on it. This is tantamount to a one-star alignment. It may not work with all telescopes.

### 12.7.3 Equatorial mode

#### Polar alignment

The easiest way to polar-align the mount of an Autostar telescope is to aim the telescope directly away from its base, so that the declination setting circle reads 90°, and then adjust the mount to aim the telescope at the celestial pole (roughly at Polaris; see p. 45). This can be done without electrical power, and the telescope can be rotated around its R.A. axis to any convenient position.

Because two-star alignment is supported, pointing accuracy is not critically dependent on polar alignment. For casual use, polar alignment errors of several degrees can be tolerated. For photography, alignment to within 1° of the true pole is desirable.

#### Selecting polar mode

Prior to aligning the telescope, you must put the Autostar in polar mode by going to the Setup menu, then choose Telescope, Mount, Polar (be sure to press ENTER). Once made, this choice is saved permanently until you change it.

#### Polar starting position

The starting position (home position) for an Autostar telescope in equatorial mode is *right side up* (with the fork arms to the sides and the finder above them), pointing directly away from the base, as in Figure 4.7, p. 46.

ETX telescopes must start within 180° of the anticlockwise limit. To meet this requirement, loosen the R.A. brake, and swivel the telescope anticlockwise (leftward) until it reaches its stop, then rightward until it is right side up again.

If you can't get it right side up without rotating more than 180°, the ETX is mounted backward. When correctly set up, its connector panel is toward the *west*.

#### Two-star alignment

The two-star alignment process is like altazimuth mode except for one screen display:

```
Polar Align
Put the telescope in the polar Home position as descr...
```

Because it is aligned on two stars, the telescope can allow for errors in polar alignment; pointing accuracy will be good even with a polar alignment error of several degrees.

Easy alignment is like two-star alignment except that the two stars are chosen for you.

### One-star alignment

The Autostar also offers alignment on Polaris plus one star, like the LX200 (p. 146). In this mode, the telescope gives you more help achieving precise polar alignment, but its pointing accuracy is then critically dependent on the accuracy you have achieved. The best of both worlds might be to do a one-star alignment in order to align the mount as well as possible, then start over and do a two-star alignment for best pointing accuracy.

## 12.8 How to interrupt a slewing movement

If the telescope is on its way to an object and you need to stop it for some reason, perhaps to keep it from tangling cords or bumping into an obstacle, just hit any of the slewing buttons (▲ ▼ ◄ ►) while it's moving. Alignment is not lost when you do this.

## 12.9 Finding objects with the built-in catalogues

The built-in catalogues of the Autostar include not only the standard ones, but also quite a few lists of objects considered interesting in various ways.

For example, under Object, Star, the Autostar has a list of named stars, a portion of the *Smithsonian Astrophysical Observatory (SAO) Star Catalog* down to magnitude 7, and several lists of interesting double stars, variable stars, nearby stars, stars with planets, and so forth.

To get to the Messier, Caldwell, NGC, or IC catalogues, choose Object, Deep Sky, and press ▲ (that is, scroll *up*). You can also reach them by scrolling down, but in that case you will go through a long set of other lists of interesting objects.

For the whole process of finding an object and reading the Autostar's description of it, see p. 212.

## 12.10 Finding objects by coordinates

To slew to a particular right ascension and declination or altitude and azimuth, hold down MODE for 2 seconds, then release it. Use the lower ▲ and ▼ buttons to select R.A. and declination or altitude and azimuth. Then press GO TO and type in the coordinates to which you want to slew. (See p. 217.) Press MODE to get back to the menu system.

Rather than type in coordinates over and over, you can store them as a user-defined object (choose Object, User Objects).

Azimuth on the Autostar is measured from north through east in the standard way, so that south is 180° and west is 270°.

## 12.11 More precise pointing

### 12.11.1 How to sync on an object

If you suspect that the telescope is not well aligned with the celestial sphere, an easy way to improve the alignment is to *sync* (synchronize) with a known object. This helps eliminate pointing errors that have accumulated while slewing around the sky.

Find any object using the built-in catalogues and go to it, or at least select it (bring it up on the screen). Press the ENTER key for at least 2 seconds and release it. The display reads:

```
Object:
ENTER to sync
```

and the drive temporarily stops. Center the object (changing slew speed as necessary to do so) and press ENTER. The telescope assigns the coordinates of the object to its current position and resumes tracking.

### 12.11.2 High-precision mode

If you need unusually precise pointing, choose Setup, Telescope, High precision, ON, and press Enter.

(Since High precision is the last item on the Telescope menu, you can reach it in one step by scrolling up instead of down.)

In high-precision mode, whenever you go to an object, the telescope will first select a nearby star and invite you to center it. If that star happens to be behind a tree or otherwise unsuitable, press ▼ for an alternative.

### 12.11.3 Square spiral search ("box scan")

In Autostar version 21 (2.1) or later, you can initiate an automatic search of the sky by pressing GO TO a second time after arriving at an object, or by holding GO TO down for more than 2 seconds.

The telescope then moves around its starting point in a square spiral pattern until you press MODE. This is a convenient way to find an object that has ended up slightly outside the field owing to inaccurate pointing.

The size of the spiral is intended to match the field of a 26-mm eyepiece. You can alter it by changing the focal length of your telescope on the Setup, Telescope menu. For example, if you are using a 13-mm eyepiece (half of 26 mm), set the focal length to twice its true value.

The real focal length of the telescope doesn't change, of course, but there is no harm in lying to the Autostar. Remember, however, that when the focal length is entered incorrectly, Autostar calculations relating to eyepieces, magnification, and field of view will come out wrong.

### 12.11.4 Drive training and backlash adjustment

The Train Drive procedure on the Telescope menu (p. 215) should be performed when the telescope is new and every few weeks or months thereafter, or whenever pointing accuracy seems to have deteriorated. It involves sighting on a land object repeatedly as you approach it from different directions. Full instructions are displayed on the keypad.

Backlash compensation is also available, under Alt Percent and Az Percent on the same Telescope menu. When it is set to zero, there is a delay when you make a slewing movement in the opposite direction to the previous one, such as slewing south immediately after slewing north. Adjust it to reduce or eliminate the delay. If set too high, backlash compensation causes movements to begin with a sudden jerk.

## 12.12 Cables, connections, and ports

### 12.12.1 The connector panel

Figure 12.3 shows the connector panel of the ETX-90; that of the LX90 is similar. The keypad plugs into the socket labeled HBX ("hand box"), attached by an 8-pin cable exactly like the LX200 declination cable (see p. 158).

The two Aux connectors are somewhat mysterious, but the LX90 manual reveals that they connect to an accessory port module with connections for an

Figure 12.3. Meade ETX-90 connector panel. LX90 is similar.

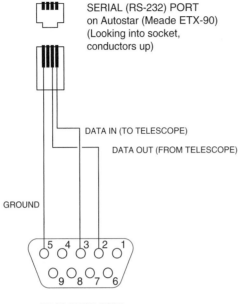

SERIAL (RS-232) PORT
on Autostar (Meade ETX-90)
(Looking into socket,
conductors up)

DATA IN (TO TELESCOPE)

DATA OUT (FROM TELESCOPE)

GROUND

TO PC SERIAL PORT

Figure 12.4. Wiring of Autostar serial (RS-232) port. The port is located on the keypad, adjacent to the socket for the keypad cable. On pin numbering see also Figure 10.9, p. 160.

illuminated reticle, an electric focuser, and a CCD autoguider like those of the LX200 (pp. 159, 161).

### 12.12.2 The serial port

Figure 12.4 shows how to connect the Autostar serial port to a personal computer for data transfer (see next section) or control by a program such as *TheSky* or *TPoint*. It is also possible to connect two Autostars together to "clone" (copy) the data from one to the other. Full instructions are included with the Meade #505 connector cable set.

The low-end Autostar 494 has different connections and uses a different cable set (Meade #506).

The Autostar instructions indicate that some Autostar telescopes (present or future?) have a 6-pin serial port like that of the LX200 (Figure 10.8, p. 160) but with only Port 1 present.

## 12.13 Upgrading the firmware and downloading data

Meade provides a PC program called Autostar Update for downloading data to the Autostar controller through the serial port (Figure 12.5). Using it to update the firmware is almost effortless – just click on one button. You do not have

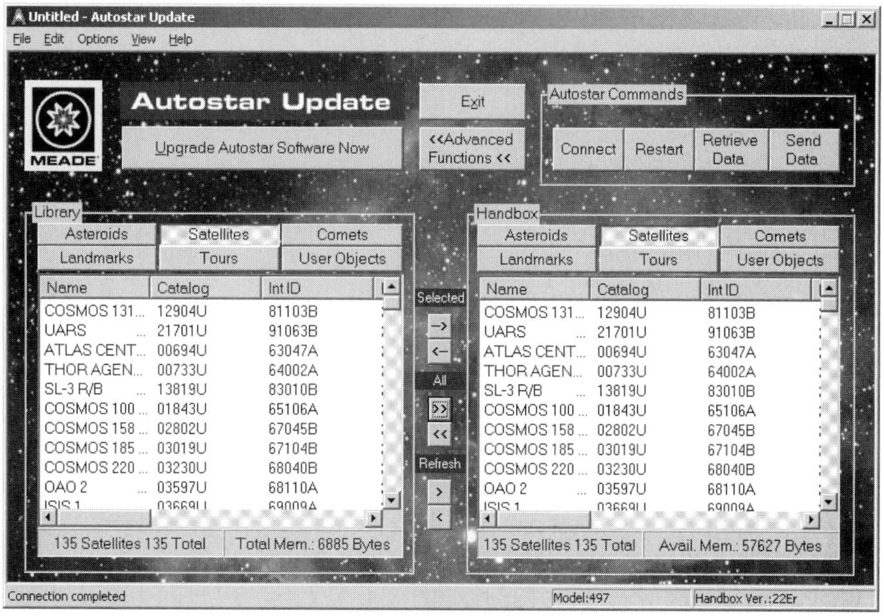

Figure 12.5. Meade provides software to update the contents of the Autostar, including its firmware. Here it is ready to download satellite orbital elements to the telescope.

to press any buttons on the Autostar before performing the update. (Earlier versions were harder to use, so in addition to downloading the latest firmware, be sure to download the latest updater!)

The same program can also download orbital elements and other information to the Autostar. These operations are more complicated but are explained in the help system. Autostar Update is available free of charge from http://www.meade.com.

## 12.14 Other advanced features

### 12.14.1 Satellite tracking

To use the Autostar to track satellites, you must download current orbital elements into the telescope. Because satellite orbits change rapidly, the orbital data should be no more than a few weeks old. Data files in standard TLE (two-line element) format are available from http://www.celestrak.com, http://oig1.gsfc.nasa.gov, and other websites. Meade's documentation indicates that updated satellite elements are available on http://www.meade.com, but during 2001, at least, this was not the case.

Once you've obtained the orbits of satellites that interest you, you can download the data into the Autostar using Meade's free software. You must also set your latitude and longitude accurately (to within $0.1°$).

Then choose `Object`, `Satellite`, select a satellite, and press ENTER:

```
Object
 Satellite
```

```
Satellite
 Select
```

```
Select Satellite
DELTA 1 R/B
COSMOS 756
COSMOS 807 R/B
 ⋮
```

The Autostar calculates whether the satellite is going to pass over your site, and if so, when. (This calculation can take a couple of minutes.) If a pass can be predicted, information is displayed:

```
COSMOS 807 R/B
Rises 05:10 AM
Sets 05:18 AM
Set Alarm
AOS Az: 142°01'
AOS Alt: 4°04'
LOS Az: 41°46'
LOS Alt: 3°16'
```

Here `AOS` and `LOS` stand for "acquisition of signal" and "loss of signal", terms borrowed from satellite radio communication. Note that most satellites pass into the Earth's shadow and are therefore invisible for part of each pass, but the Autostar is not aware of this.

Scroll down to `Set Alarm` and press ENTER. An alarm will sound one minute before the satellite's scheduled appearance. You can observe other things until then.

When the alarm sounds, return to the `Satellite` menu and select the same satellite again. Press GO TO and the telescope will slew to where the satellite will disappear from view (in order to show you where that is) and then to where it should appear. The motor drive will stop and a countdown will appear.

If the telescope happens to be aimed at a building or tree, press ENTER and the telescope will slew along the path of the satellite. Press ENTER again to stop it as soon as it's pointed at unobstructed sky.

When the countdown gets down to about 20 seconds, start watching through the telescope. When the satellite comes into view, press ENTER to start tracking it. Adjust centering with the slewing buttons.

The reason you must watch for the satellite is that the most common inaccuracy in orbital elements is that a satellite is slightly ahead of, or behind, its predicted orbit. In addition, your clock may be slightly wrong.

### 12.14.2 Sky tours

The Autostar can give tours of the sky (select Guided Tour from the main menu). Not only are several tours built-in, but others can be created and downloaded from a personal computer. Full instructions are given in the LX90 manual, and tour files are often swapped by Autostar users on the Internet. The tours are customized for what is actually in the sky at your location at the time.

Also, if you go to a constellation (under Objects), you will be offered a tour of all the bright stars in that constellation.

## 12.15 Menu maps

The following charts show the Autostar menu system as of version 22. Future versions will no doubt add features. Consult the current manual for more information.

## Autostar v. 22Er
# EXAMPLE OF FINDING A MESSIER OBJECT
Copyright 2001
Michael A. Covington

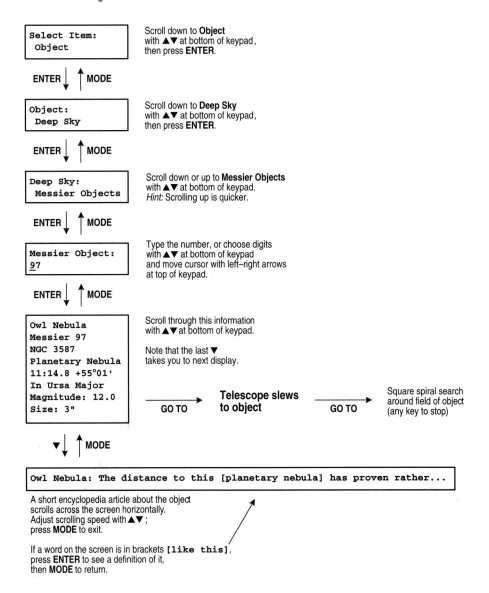

```
Select Item:
 Object
```

Scroll down to **Object**
with ▲▼ at bottom of keypad,
then press **ENTER**.

ENTER ↓   ↑ MODE

```
Object:
 Deep Sky
```

Scroll down to **Deep Sky**
with ▲▼ at bottom of keypad,
then press **ENTER**.

ENTER ↓   ↑ MODE

```
Deep Sky:
 Messier Objects
```

Scroll down or up to **Messier Objects**
with ▲▼ at bottom of keypad.
*Hint:* Scrolling up is quicker.

ENTER ↓   ↑ MODE

```
Messier Object:
 97
```

Type the number, or choose digits
with ▲▼ at bottom of keypad
and move cursor with left–right arrows
at top of keypad.

ENTER ↓   ↑ MODE

```
Owl Nebula
Messier 97
NGC 3587
Planetary Nebula
11:14.8 +55°01'
In Ursa Major
Magnitude: 12.0
Size: 3"
```

Scroll through this information
with ▲▼ at bottom of keypad.

Note that the last ▼
takes you to next display.

———→ **Telescope slews**   ———→   Square spiral search
GO TO  **to object**   GO TO   around field of object
(any key to stop)

▼ ↓   ↑ MODE

```
Owl Nebula: The distance to this [planetary nebula] has proven rather...
```

A short encyclopedia article about the object
scrolls across the screen horizontally.
Adjust scrolling speed with ▲▼ ;
press **MODE** to exit.

If a word on the screen is in brackets **[like this]**,
press **ENTER** to see a definition of it,
then **MODE** to return.

## Autostar v. 22Er
# SELECT ITEM AND OBJECT MENU MAPS
Copyright 2001
Michael A. Covington

```
Select Item:

 Setup                    See SETUP menu map.

                  ENTER
 Object          ←——————→    Object:
                  MODE        Solar System
                             Constellations      Menus to select objects
                             Deep Sky            from built-in catalogues
                             Star
                             Satellite

                             User Objects        User-defined catalogues
                             Landmarks

                                                 Identify catalogued object in or near
                                                 field of telescope. Angular distance
                             Identify            to object is displayed. Press GO TO
                                                 if you want to slew to the object.

                                                 Search catalogues for objects in the
                                                 sky of a particular size, type, and/or
                             Browse              brightness. Select Edit Parameters
                                                 to specify these characteristics, then
                                                 Start Search to begin searching,
                                                 MODE, Next to go to next item.


 Event                    Calculate times of sunrise, sunset, eclipses, other events.

 Guided Tour             Take a sky tour, either built-in or uploaded from your computer.

 Glossary                Dictionary of astronomical terms. (Quick, what is a saros?)

 Utilities               See UTILITIES menu map.
```

Select with ▲▼
at bottom of keypad.

*Hint:* **All menus wrap around at top and bottom.**
**To get to the last item on any menu, scroll *up* from the first item.**

## Autostar v. 22Er
# SETUP MENU MAP
Copyright 2001
Michael A. Covington

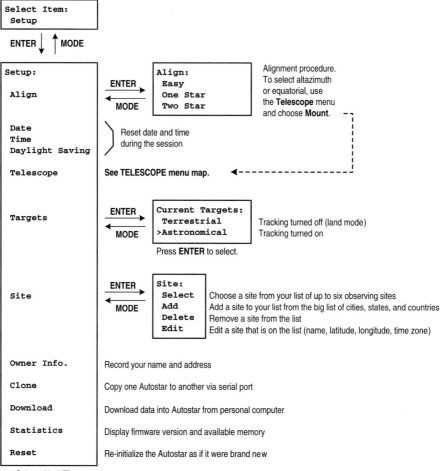

```
Select Item:
 Setup
```

ENTER ↓   ↑ MODE

```
Setup:

 Align
```
ENTER →
← MODE
```
Align:
 Easy
 One Star
 Two Star
```
Alignment procedure.
To select altazimuth
or equatorial, use
the **Telescope** menu
and choose **Mount**.

```
 Date
 Time
 Daylight Saving
```
⟩ Reset date and time
during the session

```
 Telescope
```
**See TELESCOPE menu map.** ◄-------------------

```
 Targets
```
ENTER →
← MODE
```
Current Targets:
 Terrestrial
>Astronomical
```
Tracking turned off (land mode)
Tracking turned on

Press **ENTER** to select.

```
 Site
```
ENTER →
← MODE
```
Site:
 Select
 Add
 Delete
 Edit
```
Choose a site from your list of up to six observing sites
Add a site to your list from the big list of cities, states, and countries
Remove a site from the list
Edit a site that is on the list (name, latitude, longitude, time zone)

```
 Owner Info.
```
Record your name and address

```
 Clone
```
Copy one Autostar to another via serial port

```
 Download
```
Download data into Autostar from personal computer

```
 Statistics
```
Display firmware version and available memory

```
 Reset
```
Re-initialize the Autostar as if it were brand new

Select with ▲▼
at bottom of keypad,
then press **ENTER**.

## Autostar v. 22Er
# TELESCOPE MENU MAP

Copyright 2001
Michael A. Covington

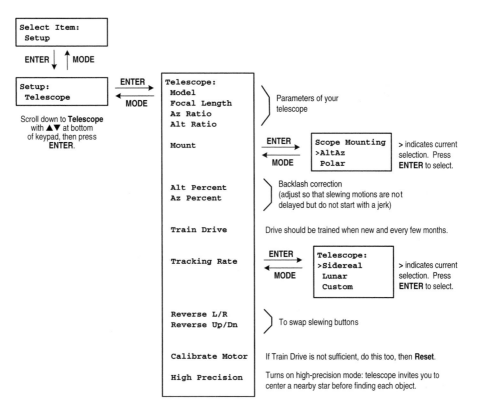

```
Select Item:
 Setup
```

ENTER ↓ ↑ MODE

```
Setup:
 Telescope
```
← ENTER →
← MODE

Scroll down to **Telescope**
with ▲▼ at bottom
of keypad, then press
**ENTER**.

```
Telescope:
 Model
 Focal Length
 Az Ratio
 Alt Ratio
```
Parameters of your
telescope

```
 Mount
```
ENTER →
← MODE
```
Scope Mounting
>AltAz
 Polar
```
> indicates current
selection. Press
**ENTER** to select.

```
 Alt Percent
 Az Percent
```
Backlash correction
(adjust so that slewing motions are not
delayed but do not start with a jerk)

```
 Train Drive
```
Drive should be trained when new and every few months.

```
 Tracking Rate
```
ENTER →
← MODE
```
Telescope:
>Sidereal
 Lunar
 Custom
```
> indicates current
selection. Press
**ENTER** to select.

```
 Reverse L/R
 Reverse Up/Dn
```
To swap slewing buttons

```
 Calibrate Motor
```
If Train Drive is not sufficient, do this too, then **Reset**.

```
 High Precision
```
Turns on high-precision mode: telescope invites you to
center a nearby star before finding each object.

## Autostar v. 22Er
# UTILITIES MENU MAP
Copyright 2001
Michael A. Covington

```
                    ENTER      ┌─────────────────┐
┌──────────────────┐   ────►   │ Utilities:      │
│ Select Item:     │           │                 │
│ Utilities        │   ◄────   │ Timer           │  ⟩ Built-in timer
└──────────────────┘   MODE    │ Alarm           │    and alarm clock
                               │                 │
   Scroll down to Utilities    │ Eyepiece Calc.  │  Calculate eyepiece magnification and field of view
        with ▲▼                │                 │
   at bottom of keypad, then   │                 │  Turn off the Sun warning and "Getting Started"
        press ENTER.           │ Display Options │  message that appear at power on
                               │                 │
                               │                 │  ⟩ Keypad display adjustments.  Adjust with ▲▼.
                               │ Brightness Adj. │    Contrast adjustment usually needed only in very
                               │ Contrast Adj.   │    cold weather.
                               │                 │
                               │ Battery Alarm   │  Beeper that sounds when battery is low
                               │                 │
                               │ Landmark Survey │  Rapid tour of all stored landmarks (defined in
                               │                 │  Object, Landmark).  Press MODE to stop.
                               │                 │
                               │ Sleep Scope     │  Turn off motors to save power without losing alignment
                               │                 │
                               │                 │  Place telescope in known starting position, so it can be
                               │ Park Scope      │  powered on again without re-aligning on the stars
                               │                 │
                               │                 │  Prevent cables from wrapping – useful if you have
                               │ Cord Wrap       │  external cables, such as a CCD camera; not needed for
                               │                 │  normal operation of ETX or LX90
                               └─────────────────┘
```

## Autostar v. 22Er
# KEYPAD DISPLAYS
Copyright 2001
Michael A. Covington

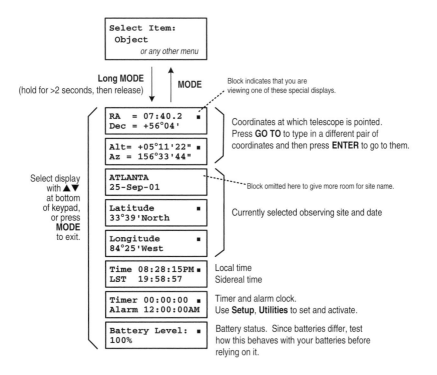

```
Select Item:
  Object
            or any other menu
```

**Long MODE** | **MODE**
(hold for >2 seconds, then release)

Block indicates that you are
viewing one of these special displays.

```
RA  = 07:40.2  ■
Dec = +56°04'
```

Coordinates at which telescope is pointed.
Press **GO TO** to type in a different pair of
coordinates and then press **ENTER** to go to them.

```
Alt= +05°11'22"  ■
Az = 156°33'44"
```

Select display
with ▲▼
at bottom
of keypad,
or press
**MODE**
to exit.

```
ATLANTA
25-Sep-01
```

Block omitted here to give more room for site name.

```
Latitude       ■
33°39'North
```

Currently selected observing site and date

```
Longitude      ■
84°25'West
```

```
Time 08:28:15PM ■
LST   19:58:57
```

Local time
Sidereal time

```
Timer 00:00:00  ■
Alarm 12:00:00AM
```

Timer and alarm clock.
Use **Setup**, **Utilities** to set and activate.

```
Battery Level: ■
100%
```

Battery status. Since batteries differ, test
how this behaves with your batteries before
relying on it.

# Index